Powder Metallurgy for Automotive and High Performance Materials in Engineering Industries

Powder Metallurgy for Automotive and High Performance Materials in Engineering Industries

P. Ramakrishnan
Professor Emeritus
IIT, Bombay
Formerly
Professor & Head of the Department of
Metallurgical Engineering
Professor of Met. Engg. & Materials Science
Chief Co-ordinator, Powder Metallurgy
and
High Temperature Materials Centre
IIT, Bombay, Mumbai

PUBLISHING FOR ONE WORLD

NEW AGE INTERNATIONAL (P) LIMITED, PUBLISHERS
New Delhi • Bangalore • Chennai • Cochin • Guwahati • Hyderabad
Jalandhar • Kolkata • Lucknow • Mumbai • Ranchi
Visit us at **www.newagepublishers.com**

Published by New Age International (P) Ltd., Publishers
First Edition: 2012
Reprint: 2015

BRANCHES

- **Bangalore** 37/10, 8th Cross (Near Hanuman Temple), Azad Nagar, Chamarajpet, Bangalore-560 018 Tel.: (080) 26756823, Telefax: 26756820, **E-mail: bangalore@newagepublishers.com**
- **Chennai** 26, Damodaran Street, T. Nagar, Chennai-600 017, Tel.: (044) 24353401 Telefax: 24351463, **E-mail: chennai@newagepublishers.com**
- **Cochin** CC-39/1016, Carrier Station Road, Ernakulam South, Cochin-682 016 Tel.: (0484) 2377303, Telefax: 4051303, **E-mail: cochin@newagepublishers.com**
- **Guwahati** Hemsen Complex, Mohd. Shah Road, Paltan Bazar, Near Starline Hotel, Guwahati-781 008 Tel.: (0361) 2513881, Telefax: 2543669, **E-mail: guwahati@newagepublishers.com**
- **Hyderabad** 105, 1st Floor, Madhiray Kaveri Tower, 3-2-19, Azam Jahi Road, Near Kumar Theater Nimboliadda, Kachiguda, Hyderabad-500 027, Tel.: (040) 24652456, Telefax: 24652457 **E-mail: hyderabad@newagepublishers.com**
- **Kolkata** RDB Chambers (Formerly Lotus Cinema) 106A, 1st Floor, S.N. Banerjee Road, Kolkata-700 014 Tel.: (033) 22273773, Telefax: 22275247, **E-mail: kolkata@newagepublishers.com**
- **Lucknow** 16-A, Jopling Road, Lucknow-226 001, Tel.: (0522) 2209578, 4045297, Telefax: 2204098 **E-mail:lucknow@newagepublishers.com**
- **Mumbai** 142C, Victor House, Ground Floor, N.M. Joshi Marg, Lower Parel, Mumbai-400 013 Tel.: (022) 24927869, Telefax: 24915415, **E-mail: mumbai@newagepublishers.com**
- **New Delhi** 22, Golden House, Daryaganj, New Delhi-110 002, Tel.: (011) 23262368, 23262370 Telefax: 43551305, **E-mail: sales@newagepublishers.com**

ISBN: 978-81-224-3031-8

₹ 495.00

C-15-06-8543

Printed in India at Glorious Printers, Delhi.
Typeset at In-house, Delhi.

PUBLISHING GLOBALLY
NEW AGE INTERNATIONAL (P) LIMITED, PUBLISHERS
7/30 A, Daryaganj, New Delhi-110002
Visit us at **www.newagepublishers.com**

Preface

The book contains selected papers from the 5th International Conference & Exhibition organized by Powder Metallurgy Association of India on High Performance Materials by Powder Metallurgy for Automotive and Engineering Industries at Goa, during Feb., 2009. These papers are arranged in four sections dealing with powder metallurgy (PM) in automotive industry, compaction techniques with improved efficiency and economy, powder plant management and special powders, and high performance meterials and processing. The first section deals with the challenges facing the PM industry, technological advances such as new powder mterial systems, new products and processes for the growth of global PM industry. Developments in sinter hardening with suitable cheaper alloying elements have resulted in products with good static and dynamic properties. Considering the engine components PM has played an important role in the development of new engines by improving the existing products by reducing the cost and by bringing newly developed systems. Investigations on PM materials and processes have resulted in the development of gears for automotive applications with comparable properties of wrought steel products. There are articles dealing with sintered metallic valve guides and metal matrix composites for brake pad applications. The section on compaction technologies covers process oriented pressing for complex parts, high speed compaction, improvements in equipment efficiencies, automated cleaning techniques for plungers in rotary presses, hot pressing resulting in improved efficiency and economy.

The open book management deals with the success story of a powder producing plant. A variety of preparation and characterization techniques have been illustrated for hafnium metal powders, spherical zirconia powders, doped nanocrystalline PZT ferroelectronics and $CaWO_4$: Ag powders for electronic display applications. The concluding section covers special materials like tungsten heavy alloys, U-plutonium carbide fuels for nuclear reactors, diamond impregnated tools and machining of PM materials and metal matrix composites.

I am thankful to the members of the Technical, Organizing, National and International Committees, Authors, Speakers, Chair persons, Sponsors, Advertisers and all those who have contributed to the success of the event and publication of this proceeding. The book is a useful and valuable resource providing comprehensive information in the rapidly expanding area of Powder Matallurgy and Particulate Materials.

P. Ramakrishnan

Contents

1

New Products and Processes and the Global Growth of Powder Metallurgy

Kalathur S. Narasimhan, Hoeganaes Corporation, U.S.A.

Abstract

Powder metallurgy has made significant advances in the conversion of wrought steel and castings in automotive applications. North America leads in the total powder metal usage in general and in particular in automotive applications. Technological advances such as advanced binder–lubrication systems, higher density processes, new material systems, improved compaction and sintering technologies contributed to this growth. In this presentation challenges facing the powder metal industry is discussed as well.

Key Words: High density, Cr containing alloys, sinterhardening, transmission market.

1.1 INTRODUCTION

The success of powder metal usage depends on a systems approach. Close collaboration among end users, parts makers and powder producers is necessary to optimize a material system for the intended application. Powder Metallurgy (PM) is well recognized as a cost effective alternate to castings, stampings, forgings and screw machining. Listed below are the advantages of PM:

- 97% material utilization
- Net shape production, eliminates or minimize machining
- Cost and energy efficient
- Can be joined with other mating surfaces
- Accepted quality, reproducibility and consistency

The markets for the powder metallurgical parts are varied. Automotive applications have so far dominated the industry growth. However, non-automotive applications are becoming increasingly important. The latter applications mainly include hand tools, household and lawn appliances and industrial motor controls and hydraulics. The market breakdown is shown in Fig. 1.1 [1]. Although most discussions of PM technology and applications focus on automotive, it will be evident that the non-automotive segment is appreciable and similarly worthy of consideration.

Automotive powder metal usage in Asia is shown in Fig. 1.2 [2]. It is anticipated that the total growth across the world will continue to increase. Auto-builds in North America have tapered off. However, the penetration of PM parts as a lower cost alternate nevertheless continues to increase and the expectation is that this trend will continue for the next several years. In addition, it is anticipated that auto-builds in China and Asia are likely to increase over the next several years. In North America, PM parts are consumed on the average of 20kg (44lbs) per car, whereas in Europe and Japan, it is estimated to be about 8–10kg (22lbs) per car. The difference in consumption is attributed to the use of automatic transmissions in North America. The growth of PM is dependent on the ability of technological advancements to deliver products with higher performance than that is currently available [3].

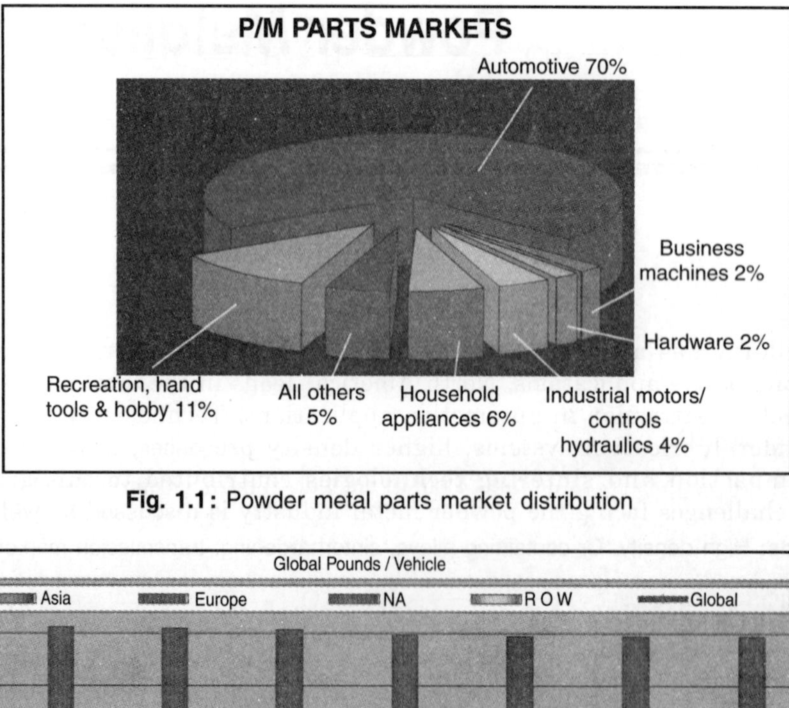

Fig. 1.1: Powder metal parts market distribution

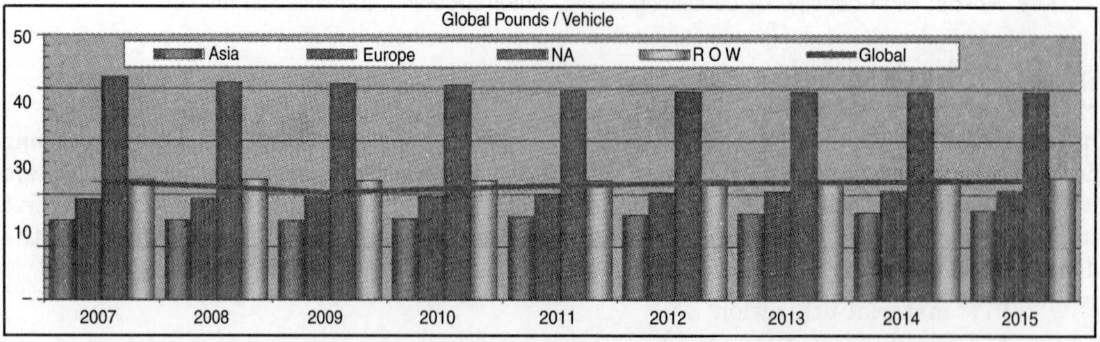

Fig. 1.2: Powder metal usage in automotive projected in Asia

We discuss below some of the new developments in technology.

1.2 ANCORSTEEL® 4300

A newly developed alloy-Ancorsteel 4300 counteracts the common oxygen-related problems associated with chromium through proprietary fabrication techniques (3). This alloy contains nominally 1.0 wt.% Cr, 0.85 wt.% Mo, and 0.6 wt.% Si, and is commercially available as a binder-treated product with nickel and graphite levels that can be tailored for specific application. The advanced processing methods allow for this alloy to be sintered at temperatures between 1120 and 1150°C. By combining chromium and silicon within one system, the alloy provides excellent strength, hardenability, fatigue, ductility, and toughness characteristics. An additional advantage of the newly developed material is that parts can be processed in conventional belt furnaces using operating temperatures that do not reduce belt lifespan. Figure 1.3 illustrates the compressibility of a new alloy in comparison with diffusion-alloyed material (FD0405) and a hybrid alloy FLN4-4405, which is prealloyed 0.85% Mo with 4% nickel and 0.6% graphite admixed.

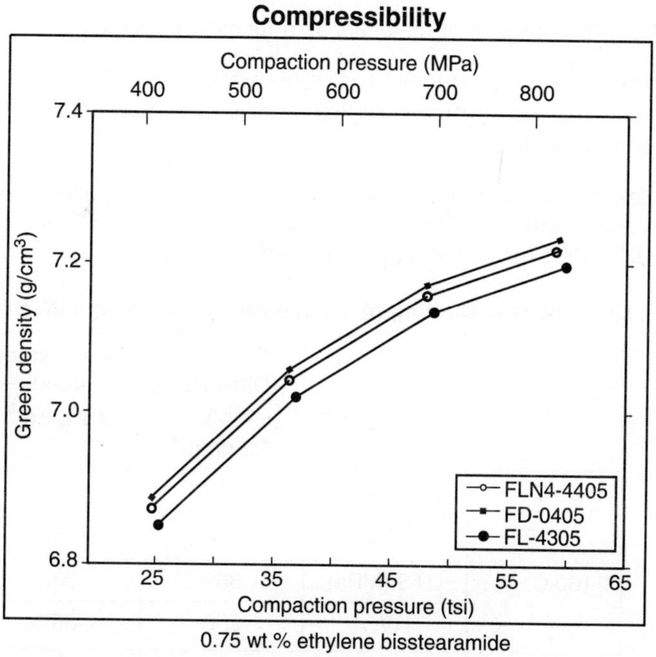

Fig. 1.3: Compressibility of Ancorsteel 4300, compared with diffusion alloyed material (FD0405) and hybrid alloy (FLN2-4405)

Figure 1.4 shows the yield strength as a function of density after sintering at 1120°C for 30 minutes in 90% Nitrogen and 10% Hydrogen atmosphere.

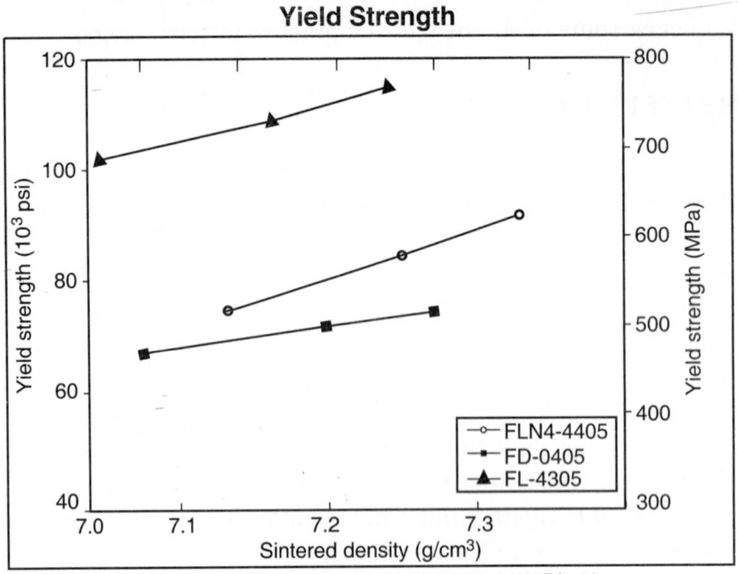

2050 °F, conventional cooling (1.3 °F/sec), tempered 400 °F for 1 h

Fig. 1.4: Yield strength of materials as a function of density shown in Fig. 1.3

Table 1.1 shows the mechanical properties with two different levels of graphite. The new alloy far exceeds the mechanical properties of a diffusion alloyed material inspite of the fact it has lower alloying elements.

Table 1.1: Mechanical properties of Ancorsteel 4300 compared to diffusion alloyed material and hybrid alloy. Sintered at Various temp. in $90N_2$-$10H_2$ atmosphere and furnace cooled (0.7°C per second) and tempered at 200°C for 1 hour. Also shown are rapid cooled (2.2 C per second) after sintering.

Conventional Cooling (Atmosphere for Sintering $90N_2/10H_2$)

		FD-4800 (Diffusion alloy) 7.2 g/cm³	Ancorsteel 4300 7.2 g/cm³
1120 C	UTS (MPa)	803	782
	HV_{10}	252	293
1160 C	UTS (MPa)	869	883
	HV_{10}	250	357
1200 C	UTS (MPa)	925	915
	HV_{10}	272	342
1250 C	UTS (MPa)	946	1047
	HV_{10}	319	401
1200 C	UTS (MPa)	1052	1096
	HV_{10}	302	375

Properties at 7.15 g/cm³

ID	YS MPa (10^3 psi)	UTS MPa (10^3 psi)	Elong %	Impact J (ft.lbf)	App. Hard HRC
4300 + 0.6 gr	760 (110)	1060 (155)	1.9	20 (15)	33
4300 + 0.4 gr	655 (95)	900 (130)	2.7	22 (16)	21
FLN4–4405	540 (78)	860 125)	2.4	23 (17)	21
FD–0405	480 (70)	860 (125)	3.2	31 (23)	95 HRB

2050 °F, conventional cooling (1.3 °F/sec), tempered 400 °F for 1 h

The dimensional stability is also superior as shown in Fig. 1.5.

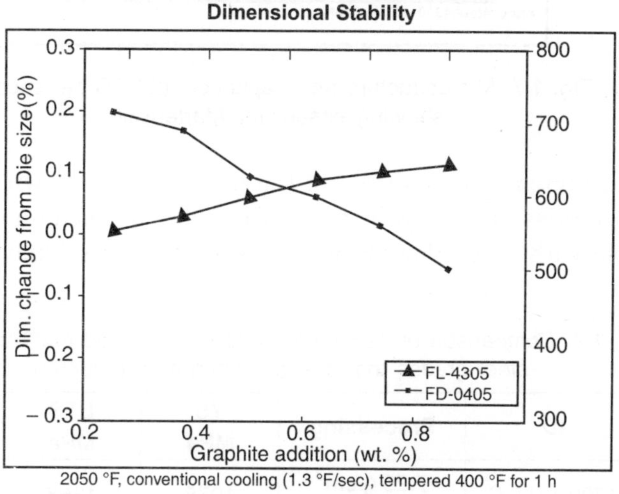

Fig. 1.5: Dimensional stability, as shown by the dimensional change as a function of graphite content. Dimensional change from die was measured after sintering at 1120°C and tempering at 200°C for 1 hour

Figures 1.6 and 1.7 show the benefits of accelerated cooling after sintering. At faster cooling rates more martensite is formed which reflects in higher strength and hardness.

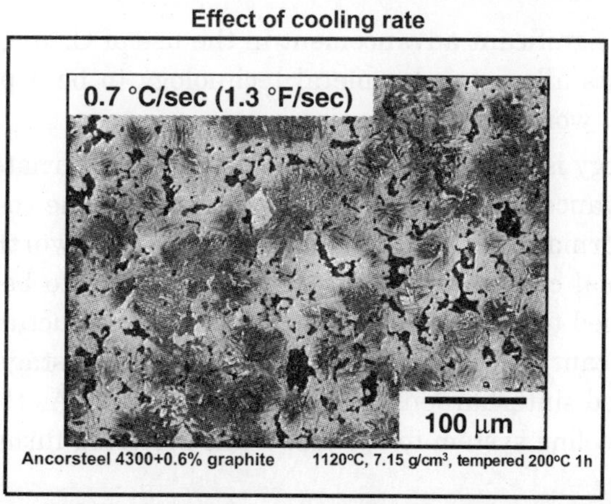

Fig. 1.6: Microstructure of 4300 grade after conventional cooling after sintering and tempering at 200°C 1 hour. Mixture of Martensite and pearlite and Bainite

Effect of cooling rate

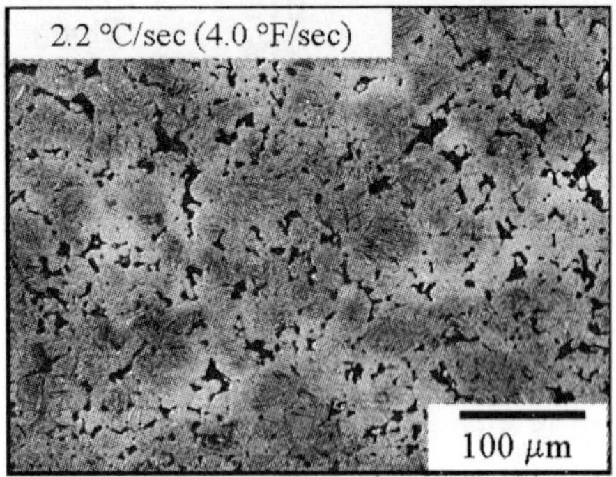

Fig. 1.7: Microstructure after rapid cool (2.2 °C per second) showing essentially Martensitic

In Table 1.2, a comparison is shown of the new alloy grade with wrought steel grades. At 7.0g/cm^3 the properties are equivalent to AISI 8620 quenched and tempered. With higher densities achievable with new alloy grade the performance is likely to exceed the wrought steel grades.

Table 1.2: Comparison of as sintered and rapid cooled 4300 with wrought steel grades that are quenched and tempered.

ID	Processing	YS MPa	UTS MPa	Hard HV10
4300 + 0.5 C	7.25 g/cm^3 2.2 °C/sec	1075	1365	440
AISI 4340*	Wrought Q&T	1260	1725	510
AISI 8620*	Wrought Q&T	965	1325	450

AISI 4340 = DIN 40 NiCrMo6 and JIS SNCM 439

AISI 8620 = DIN 20 NiCrMo2 and JIS SNCM 220H

*AISI volume 1. Properties and Selection

HOEGANAES CORPORATION

This new alloy is a significant advancement in the use of Cr in powder metal parts. Such continuous advancements allow powder metal technology to be a preferred as a lower cost alternate to other metal working technologies.

As powder metallurgy is being accepted as a lower cost alternate to other metal working technologies further advancements were necessary to reduce the cost of processing and also improve material performance. Two developments are noteworthy of mentioning. Heat treatment adds additional cost, as parts after sintering have to be heated to austenitizing temperature and quenched to generate the martensitic microstructure. Often, heat treatment process produces significant distortions leading to material wastage. In order to avoid heat treatment a process called sinterhardening (4) became popular. As the furnace manufacturers were able to add rapid cooling system to the end of the sintering furnace, this allowed sintered parts exiting the furnace to be cooled directly from the austenite phase to predominantly to martensitic microstructure.

Table 1.3 lists the standard sinterhardening grades. Also Hoeganaes recently developed a sinterhardening grade called Ancorsteel 737SH that allows parts producers to achieve hardness equivalent to heat treated parts in large sections.

Table 1.3: Typical sinterhardening grades before the development of Ancorsteel 737SH and Ancorsteel 4300

Standard Sinterhardening Materials

Sample Chemistries: MPIF 35 designations

Alloy	Mo (%)	Cu (%)	Ni (%)	C (%)	Alloy Type
FLN2-4408	0.85	-	2.0	0.8	Hybrid
FLN4-4408	0.85	0	4.0	0.8	Hybrid
FLN6-4408	0.85	-	6.0	0.8	Hybrid
FLN-4608	0.50	-	4.0	0.8	Hybrid
FLC-4608	0.50	2.0	2.0	0.8	Hybrid
FLC-4908	1.50	2.0	-	0.8	Hybrid
FLNC-4408	0.85	2.0	2.0	0.8	Hybrid
FD-0208	0.5	1.5	1.7	0.8	Diff. Bonded
FD-0408	0.5	1.5	4.0	0.8	Diff. Bonded

1.3 ANCORSTEEL 737SH CAPABILITIES

A mix of 737SH, 2% Cu, and 0.9% graphite (coded 1.4 Ni-1.25 Mo-0.4 Mn-2 Cu-0.8 C FLC-24808) was compared to two common sinter-hardening grades used in North America, denoted by MPIF material codes FLNC-4408 and FLC-4608 as shown in Table 1.4.

Table 1.4: Nominal alloy compositions (%)

ID	Fe	Mo	Ni	Mn	Cu	Gr
FLCN-4408	Bal.	0.8	2.0	0.1	2.0	0.9
FLC-4608	Bal.	0.5	1.8	0.1	2.0	0.9
737SH+2Cu+0.9gr	Bal.	1.3	1.4	0.4	2.0	0.9

Figure 1.8 shows comparative microstructures of the alloys. FLC-24808 is clearly more hardenable than the other two alloys, with a 100% martensitic microstructure under these processing conditions.

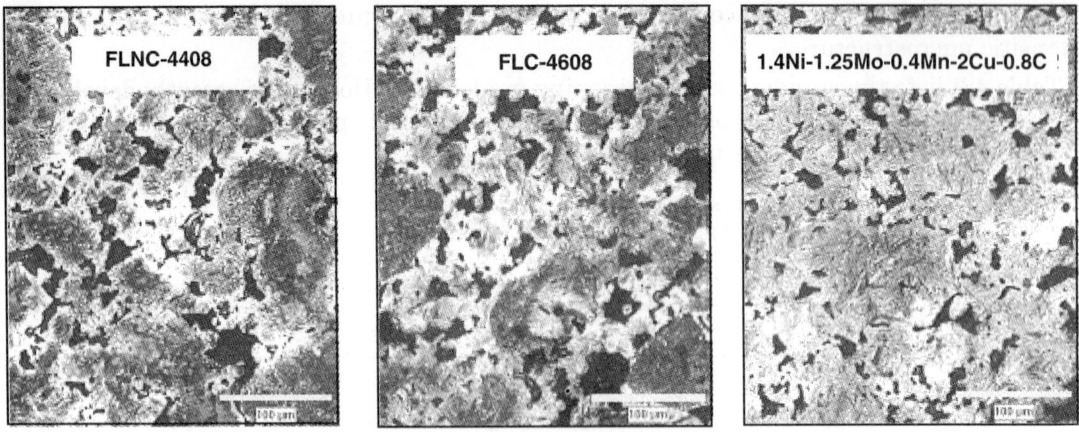

Fig. 1.8: Comparison of conventional sinterhardening grades developed earlier with Ancorsteel 737SH

These microstructural differences lead to significant changes in mechanical properties. The tensile strength of the FLC-24808 mix has a far superior strength level compared to FLNC-4408 at the same or slightly lower densities, as shown in Fig. 1.9.

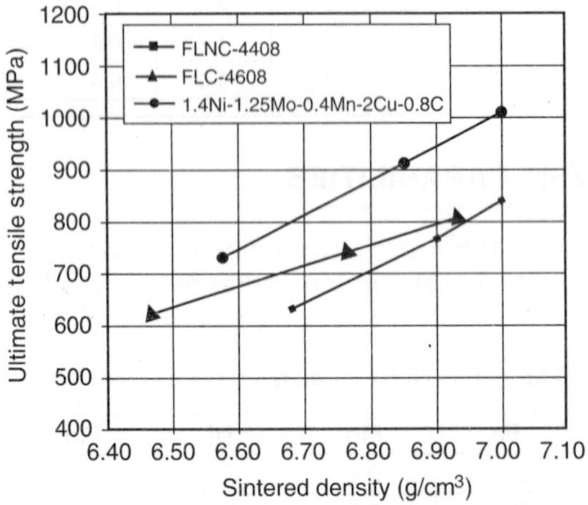

Fig. 1.9: Tensile strength of 737SH with other sinterhardening grades developed earlier at different densities

This is a direct result of the complete transformation of the microstructure to martensite. The same trend is observed with FLC-4608. Apparent hardness increases through complete transformation of the microstructure to martensite and by higher sintered densities. Figure 1.10 shows how the superior compressibility and hardenability of the FLC-24808 mix results in higher apparent hardness values. A much higher cooling rate would be required to raise the hardness of the other materials to the same level.

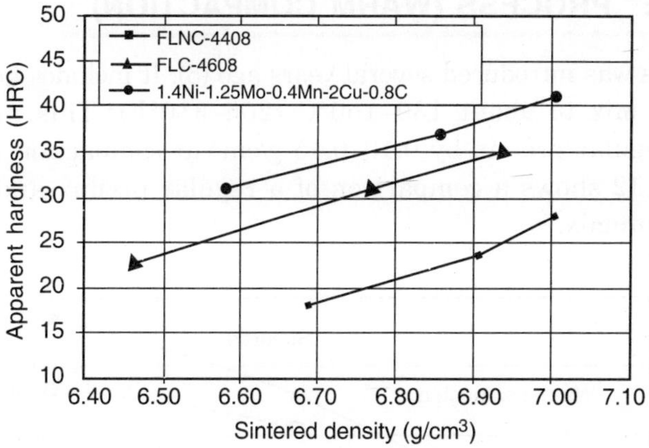

Fig. 1.10: Apparent hardness of 737SH compared with other sinterhardening grades developed earlier

1.4 HIGHER DENSITY

The continuous quest in the PM industry for higher performance requires higher density. Higher density can be achieved by using higher compaction pressures and higher compaction temperatures. However there exists a natural limit to reaching higher densities, the pore-free density. Pore-free density is the density of a green compact if all the porosity could be removed. Pore-free density is dependent on the specific gravity and the per cent of each premix addition. Low specific gravity ingredients like graphite and lubricant/binder additions have significant negative effects on pore-free density. Figure 1.11 shows the effect of lubricant addition on the pore-free density of a FN-0208 premix. In most real life applications the powder can be compacted only to about 98% of the pore-free density.

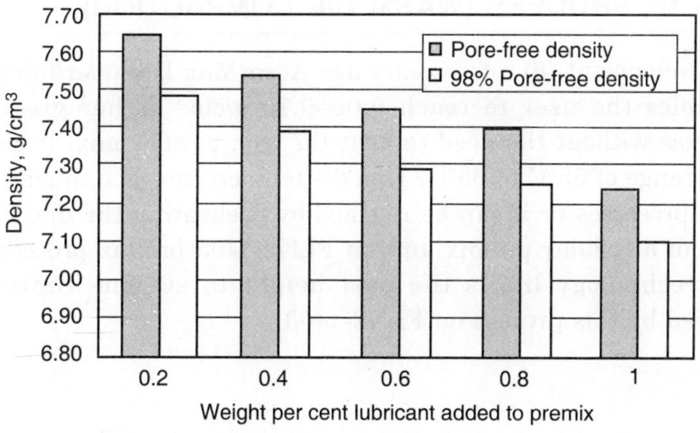

Fig. 1.11: Pore-free density of 0.85 Mo Prealloy + 2% nickel + 0.6 graphite calculated from the specific gravity of the ingredients and when this premix is compacted to remove all porosity inside die cavity. The pore-free density is shown as a function of added lubricant content in the premix

1.5 ANCORDENSE® PROCESS (WARM COMPACTION)

ANCORDENSE process was introduced several years ago (5). It includes heating both the tooling and the iron powder mix to about 130–150°C (265–310°F). This process increased the compressibility of a regular premix by 0.10–0.15 g/cm^3 at compaction pressures at least up to 700 MPa. Figure 1.12 shows a comparison of a regular premix and a warm compaction processed FLN2-4405 premix.

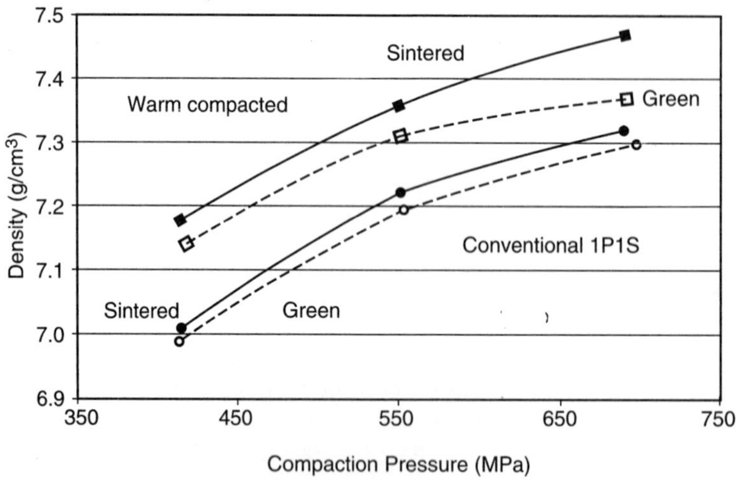

Fig. 1.12: Green and sintered density obtained by Ancordense warm compaction process compared with conventional compaction as a function of compaction pressure. 85 HP + 2% nickel + 0.6% graphite + lubricant

1.6 ANCORMAX D® PROCESS (WARM DIE COMPACTION)

In a more recent development (6) a few years ago AncorMax D binder/lubricant system was introduced. It enables the user to reach 0.05–0.15 g/cm^3 higher green densities than conventional premixes without the need to heat the iron powder mix. It only requires a die temperature in the range of 55°C to 65°C. This die temperature is frequently reached during regular compaction processes or it can be reached by preheating the die. Figure 1.13 shows the compressibility of a regular premix and an FLN2-4405 premix processed via warm die compaction. This technology limits the part height to 20 mm. Sintered densities of 7.45 g/cm^3 is achieved by this process on FLN2-4405.

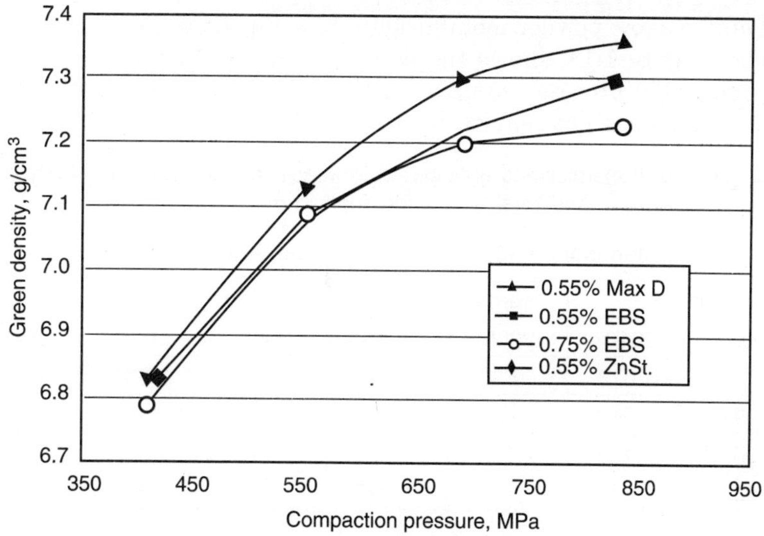

Fig. 1.13: Green density of warm die compacted part (AncorMax D Process) compared with conventional lubricants used in the industry. (Ethylene bis-stearmide; Zinc Stearate)

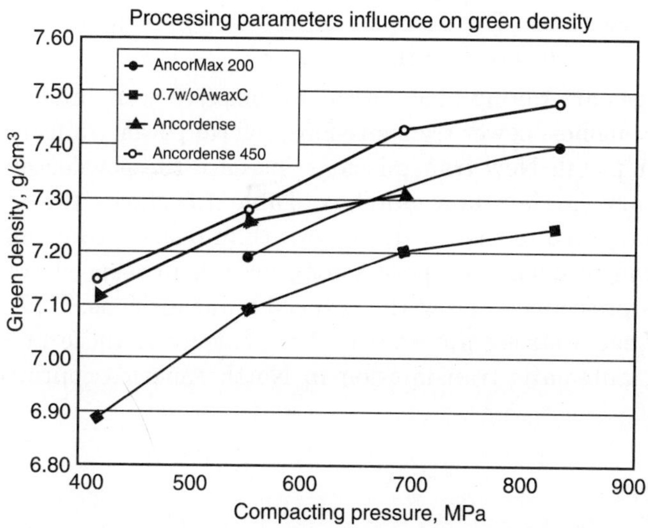

Fig. 1.14: Green density of Ancorsteel 85HP + 2% Ni + 0.35% graphite processed by various high density processes

1.7 ANCORMAX 200 PROCESS

This newer technology (7) reduces the lubricant further to 0.4% form 0.55% used for AncorMax D. Allows complex shapes can be produced. This requires tools to be heated to 83°C and requires no powder heating. Parts as high as 40 mm can be produced. Densities achieved are close to 7.5g/cm^3 after sintering at 1120°C.

All these advances allow powder metallurgy to be competitive and meet growing demands from the end users. Figure 1.14, shows the density achieved by these advanced compaction processes. Ancordense 450 process requires heating of the powder and the die as compared to Ancormax 200 which does not require heating of the powder.

Table 1.5: Type of transmissions anticipated in usage in Asia, Europe, North America and total worldwide (unit in millions)

Transmissions (000), Type in Europe, Asia and Nafta								
Region	**Manual**		**Semi-Automatic**		**CVT**		**Automatic**	
	2005	2010	2005	2010	2005	2010	2005	2010
Worldwide	27,786	31,673	630	1373	2,415	3,391	26,077	26,806
Asia	11,094	13,563	53	65	1,946	2,551	12,528	13,481
Europe	16.054	17,493	567	1,271	265	296	2,559	1,792
NAFTA	637	616	9	37	202	543	10,989	11,532

PM parts manufactured today utilize several of the material systems described above. Any specific application is dependent on design and material performance to meet the design needs. For the same application, there could be more than one material system and process to match the performance characteristics. The exact material selection is made by discussions with the powder producers, parts producers, and the end user.

Number of changes are taking place in the automotive scene and these may affect the growth of PM. Newer engines, newer transmissions, hybrid power trains all impact differently the growth of powder metal. New transmissions increase the powder metal usage, as they require more planetary carriers and clutches. Table 1.5 shows the transmission builds worldwide. NAFTA projected to show a strong growth in 6-speed transmission and moderate growth in 5-speed transmission, Europe a strong growth in 6-speed transmission as well. Asian market also is projected to grow the 6-speed transmissions. CVT make inroads into NAFTA market, but has a strong presence in Asia. Inspite of these manual transmissions in Europe and Asia, automatic transmission in North America continues to show strong growth.

1.8 CONCLUSIONS

Growth of powder metal parts is fueled by technological advancements. Newer chromium containing grades offer easier processing and offer opportunities for replacing number of wrought steel grades such as AISI 8620 (JIS SNCM 220H). Sinterhardening grades offer a lower cost option to replace heat-treated parts. Higher density compaction technologies offer lower cost to replace double press double sinter operations and achieve densities approaching 7.5g/cm^3. Newer transmissions increase powder metal usage whereas hybrids and CVT reduces the powder metal components usage.

REFERENCES

1. K.S. Narasimhan, "*Recent Advances in Ferrous Powder Metallurgy, AdvancedPerformance Materials 3*", pp. 7–27 (1996).
2. Source–Automotive Magazines–Data compiled by Eric Boreczky (Private Communication)
3. P. King, and B. Lindsley, "*A New Cr-Bearing Alloy for High Performance Applications*", Advances in Powder Metallurgy & Particulate Materials, MPIF, Princeton, NJ 2005.
4. M.C. Baran, et al., "*A Superior Sinter-Hardenable Materials*", Advances in Powder Metallurgy & Particulate Materials, MPIF, Princeton, NJ 1999.
5. F.G. Hanejko, S.H. Luk, H.G. Rutz, "*A New Method For Manufacturing High Performance Powder Metallurgy Components*", Advances in Powder Metallurgy & Particulate Materials 1994, Vol. 5, Compiled by C. Lall and A. J. Newpaver, pp. 135–163, MPIF, Princeton, NJ.
6. G. Poszmik and M. Marucci, "*Higher Density and Higher Performance By Single Pressing and Sintering*", International Conference on Powder Metallurgy and Particulate Materials, Chicago, IL Jun. 13–17, 2004.
7. F.G. Hanejko, Presentation at International Conference on Powder Metallurgy and Particulate Materials, Chicago, IL Jun. 13–17, 2004.

❍ ❍

2

An Industry Overview on Sinter Hardening Solutions

Ulf Engström and Robert Frykholm, Höganäs AB, Sweden

Abstract

Sinter hardening of PM components is a well established and cost effective alternative to secondary heat treatment as one step in the manufacturing chain can be eliminated. However, during the last years alloys suitable for sinter hardening have been facing increased cost pressure due to increases and fluctuations in prices of nickel and molybdenum. Therefore, there is an increasing interest for availability of more cost effective low alloy steel powders that realize a compromise between compressibility and hardenability. The purpose of this paper is to present both static and dynamic properties achievable with a number of cost effective low alloy steel powders after sinter hardening.

2.1 INTRODUCTION

The development of new materials and cost effective processes are two main factors that contributed to the growth of PM during the last decades. Sinter-hardening, which utilises high performance materials in combination with an accelerated post sintering cooling rate, is one example of such material and process development (1,2,3). As this technique is a one step process it combines good manufacturing economy with the ability to achieve a unique combination of mechanical properties.By controlling the post sintering cooling rate, the microstructure can be manipulated to form the required amount of martensite to obtain the desired mechanical properties. By increasing the hardenability of the material large amounts of martensite can be still obtained at lower cooling rates. Alloying elements such as molybdenum, chromium, nickel and copper all promote hardenability. In recent years there has been a dramatic rise in raw material costs of especially molybdenum, nickel and copper. These are typical alloying elements used in applications utilizing traditional sinter and sinter hardened PM materials. Therefore there is a continuously growing interest to utilize more cost effective alloying elements such as chromium and leaner nickel/molybdenum alloyed steels.

In this paper the influence of alloying system and cooling rate on both microstructure and mechanical properties will be discussed. Besides composition and microstructure, density and mass play an important role to achieve the desired properties. Therefore properties after sinter hardening were evaluated at two density levels, 7.0 glcm^3 and 7.15 glcm^3. The influence of mass on the resulting microstructure and hardness after sinter hardening will also be discussed.

2.2 EXPERIMENTAL PROCEDURE

Materials

In order to compare the sinter hardening performance of two pre-alloyed Cr alloyed PM steels and two new cost effective Ni/Mo steels with a conventional Ni/Mo steel, Astaloy A, five different powder mixes were prepared. The base powders used for the mixes were Astaloy A , Astaloy CrL, Astaloy CrM, Astaloy LH and Distaloy LH. The chemical composition of the different powder mixes is shown in Table 2.1 below. Astaloy A is a fully pre-alloyed powder with 1,9% Ni and 0,5% Mo. Astaloy CrL is a fully pre-alloyed powder containing 1,5% Cr and 0,2% Mo whereas Astaloy CrM is pre-alloyed with 3% Cr and 0,5% Mo. Astaloy LH is a new fully pre-alloyed powder with 0,9% Ni and 0,9% Mo. The Distaloy.LH powder is based on the fully pre-alloyed Astaloy LH to which 2% Cu has been diffusion alloyed. The identification codes for these mixes used in this paper are also shown in this table. 0.60 % Kenolube was added as lubricant. The copper admixed to the different was -325 mesh electrolytic copper powder, and the graphite used in all cases was UF-4 from Kropfmühl.

Table 2.1: Chemical composition of investigated powder mixes

Base powder	Identification code	Ni %	Mo %	Cr %	1 Cu %	Gr. %	Lub. %
Astaloy CrL	CrL		0,2	1,5	2,0**	0,85	0,6
Astaloy CrM	CrM		0,5	3,0		0,45	0,6
Astaloy A	AA	1,9	0,5		2,0**	0,85	0,6
Astaloy LH	ALH	0,9	0,9		2,0**	0,85	0,6
Distaloy LH	DLH	0,9	0,9		2,0*	0,85	0,6

*Diffusion bonded ** Admixed

Processing Conditions

Test specimens, tensile test bars, impact energy bars and plane bending fatigue specimens according to ISO standards were compacted in a Tonitechnik 100 ton, semi-automatic laboratory press, to two different density levels, 7.00 g/cm^3 and 7.15 g/cm^{3}by conventional compaction. It is important to notice that the compaction pressure required to reach the desired density levels will vary considerably with the base material selected.

Sintering was carried out in a laboratory sintering furnace from Cremer equipped with rapid post sintering cooling system to be able to accelerate the cooling capabilities of the continuous belt furnace to the desired cooling rate. The sintering temperature used was set to 1120 °C. Sintering time was 30 min at temperature and the atmosphere used was 90/10 mix of nitrogen and hydrogen. Two different cooling rates were used, 0.8 °C/s for reference and 2.5 °C/s for sinter hardening. After sintering tempering was carried out at 200 °C for 60 minutes in air.

After the final tempering operation, apparent hardness measurements were performed on the surface of the specimens using the Rockwell C scale for comparison purposes. Tensile testing was performed in a MTS Sintech 20D tensile testing machine. Impact energy tests were carried out in a Charpy tester. Plane bending fatigue tests were carried out on un-notched compacted specimens at a fully reversed load ratio of R = –1. Endurance limit was evaluated at 2 million cycles.

Hardenability

Alloy composition, carbon content and cooling rate have a strong influence on the final microstructure of the material. By utilizing powders alloyed with chromium, nickel, molybdenum and copper together with proper amounts of carbon completely martensitic structures can be formed after sintering at 1120 °C already at moderate cooling rates, 1.5~2.5 °C/s.

To investigate the hardenability specimens were pressed to 7.0 g/cm^3 and sintered in $90N_2/10H_2$ atmosphere for 20 minutes at 1120 °C. They were then held at 1120 °C in vacuum in the dilatometer for 10 minutes before they were cooled at well defined cooling rates down to room temperature. Fractions of martensite were determined by light optical microscopy of polished and etched sample cross sections.

Figure 2.1 shows the amount of martensite obtained for the DLH and AA materials using different cooling rates during sintering. The hardenability behaviour is similar for the two materials. The AA material has a somewhat better response at slow cooling rates (>0.5 °C/s), while DLH ensures higher levels of martensite at cooling rates in the range 1.5-3 °C/s. Figure 2.2 shows below shows micrographs of etched structures in DLH. In the micrographs the bright regions are martensite, and the dark brown regions are bainite. The images clearly show that high levels of martensite can be obtained at moderate sinter hardening cooling rates (~1.5 °C/s).

In Figure 2.3 hardenability curves for the CrL material is presented. As can be seen cooling rates >2 °C/s are necessary, otherwise pearlitic/ bainitic structures are obtained. If high martensite contents are requested also for slower cooling rates, Cu can be admixed. This significantly improves the response at low cooling rates.

Figure 2.4 shows microstructures obtained for CrL with and without an addition of 2% Cu after sinter hardening with a cooling rate of 2.5 °C/s. As can be seen the microstructure in the specimen without Cu is pearlitic/bainitic while the Cu containing material is mainly martensitic with some presence of bainite.

The more high alloyed CrM shows good hardenability already at 0.4% graphite, Figure 2.5 below. At 0.5% graphite the hardenability is excellent with a stable martensitic plateau from a cooling rate of 1 °C/s. At lower cooling rates there will be some fine bainite present. The high stability in microstructure makes CrM a suitable choice when sinter hardening parts with both thin and more massive sections, and a requirement for high strength.

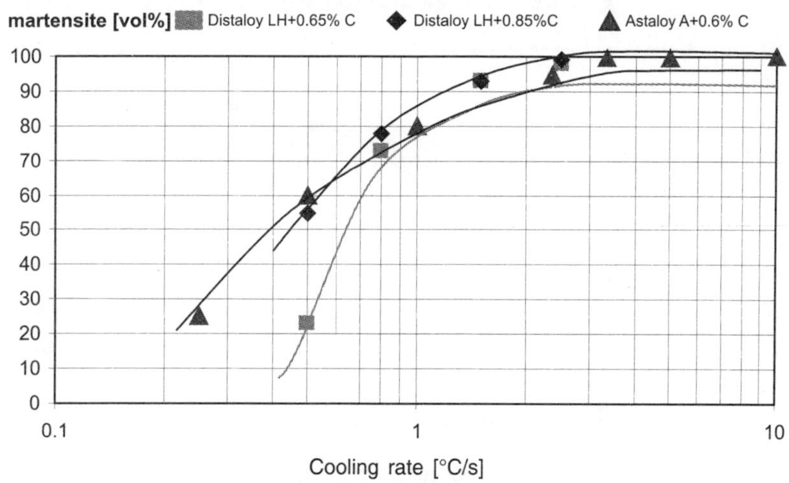

Fig. 2.1: Martensite content vs. cooling rate for Distaloy LH+0.65% and 0.85% graphite and Astaloy A+2%Cu+0.65% graphite

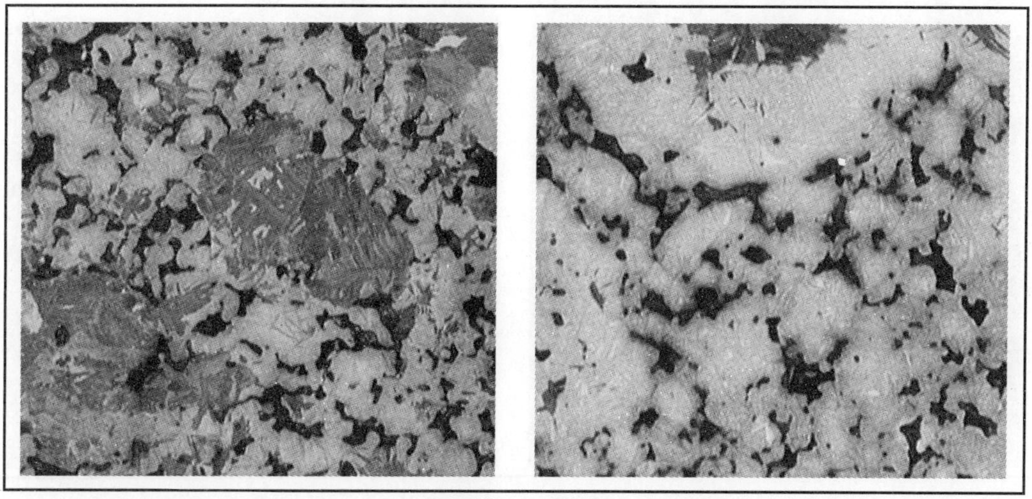

Magnification 300X

Distaloy LH+0.77C, 0.5 °C/s Distaloy LH+0.78C, 1.5 °C/s

Fig. 2.2: Microstructure of Distaloy LH after different cooling rates; 0.5°C/s and 1.5°C/s

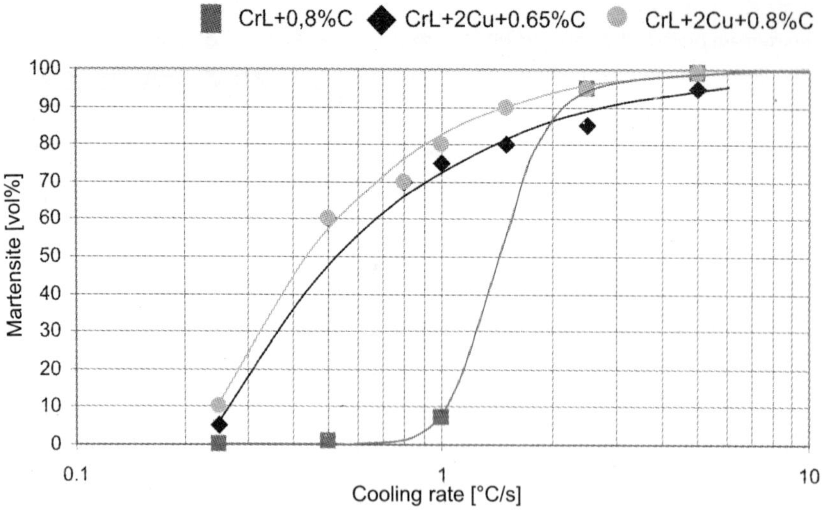

Fig. 2.3: Martensite content vs. cooling rate for Astaloy CrL with and without 2%Cu

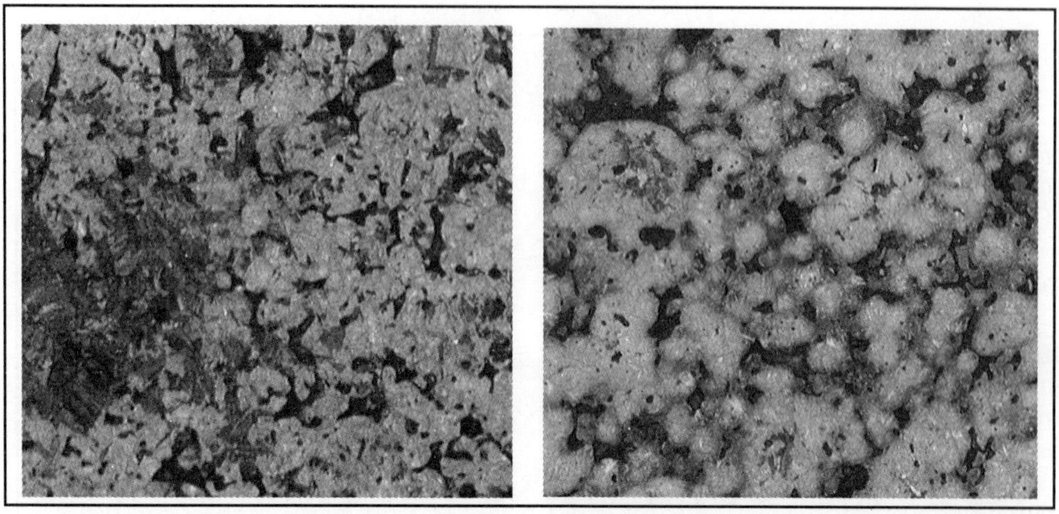

Magnification 300X

Astaloy CrL+0.69C, 2.5 °C/s Astaloy CrL+2Cu+0.71C, 2.5 °C/s

Fig. 2.4: Astaloy CrL + 0.70% C with and without 2% Cu. Cooling rate; 2.5°C/s

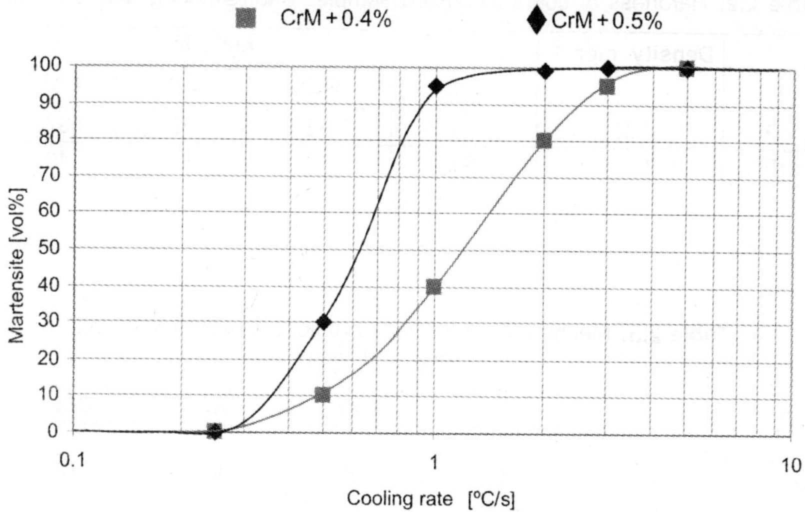

Fig. 2.5: Martensite content vs. cooling rate for Astaloy CrM with 0.4% C and 0.5% C

2.3 RESULTS AND DISCUSSION

A comparison of the hardness levels obtained at the two different density levels is shown in Table 2.2. As can be seen the hardness has increased for all materials when the density is increased from 7.00 to 7.15 g/cm^3 which is to be expected. The hardness levels obtained for the CrM material at the lower density level is the same as for the CrL material and slightly lower compared to the other materials. This is mainly due to the lower carbon content, 0.45%. However, at the higher density level the CrM material shows comparable hardness to the DLH and ALH materials.

As can be seen all properties increased with increasing density. The ultimate tensile strength increased about 10% for the two Cr alloyed and the DLH and ALH materials when density was increased from 7.00 to 7.15 g/cm^3. The CrM material and the DLH material achieved the same tensile strength at both density levels, 955 and 1060 MPa respectively. The CrL and AA material achieved comparable results at both density levels and about 10% lower compared to the other three materials. It is also worth to point out that the the carbon content of the CrM material was 0.45% compared to 0.85% for the other materials and without any Cu addition.

The increase in density resulted in improvements in yield strength of about 10% for all materials. The CrM material obtained the highest yield strength at both density levels, 928 and 1020 MPa respectively. All the other materials obtained similar yield strength at the two density levels, 800 to 840 MPa at the lower level and 905 to 935 at the higher density level.

By comparing the impact energies obtained for the five materials it can be seen that the DLH and ALH materials achieved the highest level, 17 J at the lower density and 20 J at the higher density. The other three materials obtained slightly lower impact energy at both densities.

Table 2.2: Hardness of cold compacted samples after sintering and tempering

	Density, g/cm^3	Material				
		CrL	CrM	AA	DLH	ALH
Hardness, HRC	7.00	33	33	35	37	35
Hardness, HRC	7.15	34	40	37	40	41

The mechanical properties obtained after sinter hardening of the five materials are shown in Table 2.3.

Table 2.3: Mechanical properties of cold compacted samples after sintering and tempering

Material	CrL	CrM	AA	DLH	ALH
Density 7.00 g/cm^3					
Tensile strength, MPa	850	955	889	955	925
Yield strength, MPa	800	928	815	825	840
Impact energy, J	16	15	16	17	17
Bending fatigue strength, Mpa	290	320	270	295	290
Density 7.15 g/cm^3					
Tensile strength, MPa	935	1060	940	1060	1015
Yield strength, MPa	905	1020	910	908	935
Impact energy, J	19	18	19	20	20

The increased cooling rate also has a positive effect on plane bending fatigue strength. CrM obtained the highest fatigue strength, 320 MPa at the lower density level. The CrL and LH based materials obtained similar fatigue strength levels, 290–295 MPa compared to 270 MPa for material A. It should be pointed out that bending fatigue strength levels could be further increased by combining increased cooling rate with sintering at higher temperature.

The dimensional change obtained for the five materials during sintering is shown in Table 2.4. As can be seen the materials containing Cu all show a swelling of about 0.2–0.3% , which is due to alloying with Cu which promotes growth during sintering. Regarding CrM it shows a slight shrinkage during sintering with a dimensional change close to zero.

Table 2.4: Dimensional change for the different materials at density 7.00 g/cm^3

Material	CrL	CrM	AA	DLH	ALH
DC, %	0,23	– 0,02	0,39	0,32	0,33

2.4 INFLUENCE OF SIZE AND MASS

Slugs with an outer diameter of 40 mm and varying heights, 10, 15, 20 and 30 mm, were compacted to simulate the different sizes of components manufactured in the PM industry. In this study the DLH mix outlined in Table 2.1 was compared with a mix containing 0,65%C and 1,5% Cu in order to see the influence of carbon, copper and mass effect on the resulting microstructure. The slugs of the two powder mixes were compacted to a green density of

7.00 g/cm³. Sintering was performed for 30 minutes at 1120°C in a 90/10 N_2/H_2 atmosphere. The cooling rate of the furnace was set to correspond to 1.5°C/s for the smallest slugs. The slugs were then tempered at 200°C for 1 hour in air. Apparent hardness and microstructure were then evaluated for each slug size.

The results are summarized in Table 2.5 below. As can be seen the DLH material obtained a hardness level exceeding 35 HRC when the mass increased from 88 g to 264 g. For the other material containing 1,5%Cu and 0,65%C the hardness decreased continuously with increasing mass. By comparing the microstructures achieved for the two materials when the mass is increasing, it is clear that the amount of martensite formed has decreased significantly with increasing mass for the material with the lower amount of Cu and C. The martensite content is reduced from almost 100% for the smallest slug to an almost completely bainitic microstructure for the largest slug. The DLH material on the hand showed a mainly martensitic structure for all slug sizes.

Table 2.5: Surface hardness and vol. % martensite after sinter hardening of slugs with 40 mm diameter with different heights and weights

	Slug size (0.DxH),mm	40 x 10	40 x 15	40 x 20	40 x 30
	Weight, g	88	126	176	288
Material	Eff. Cooling rate, °C/s	1.5	1.3	1.2	1.0
DLH (1.5%Cu)+0.65%C	Vol. % Martensite	90–93	70–75	65–70	2–5
	Hardness, HRC	34	31	30	10
DLH+0.85%C	Vol. % Martensite	98	97	95	85
	Hardness, HRC	37	37	37	36

Comparison of cost and performance for different PM steels

The cost effectiveness of the pre-alloyed Cr materials relative to other sinter hardening grades maes these materials very competitive for new high performance sinter hardening applications. Table 2.6 clearly demonstrates the superior cost effectiveness of the CrL and CrM materials. Using DLH as a reference, the material cost for CrL and CrM can be up to 30% lower.

Table 2.6: Relative material cost index for the different PM materials, Quarter 1, 2008

Material	CrL	CrM	AA	DLH	ALH
Relative cost, %	67	74	103	100	98

2.5 CONCLUSIONS

The results presented in this paper show that the Cr-Mo pre-alloyed materials Astaloy CrL and CrM on one hand and the newly developed materials Distaloy LH and Astaloy LH on the other are suitable for sinter hardening applications. Both LH materials are cost effective sinter hardening solutions with high compressibility that can be sintered in any sintering atmosphere, including endothermic gas.

Astaloy CrM and Astaloy CrL are the most cost effective sinter hardening solutions when the sintering process can handle sintering of Cr in a proper way.

REFERENCES

1. Akpan, E., L' Esperance, G., Roy, L., "*Case Histories with Sinter Hardening Low Alloy Steel Powder*", Advances in Powder Metallurgy & Particulate Materials–1993, Vol. 4, p. 289, Metal Powder Industries Federation, Princeton, NJ.
2. Höganäs Handbook for Sintered Components & Metallography. 1999, p. 197.
3. U. Engström, "*Evaluation of sinterhardening of different PM materials*", Advances in Powder metallurgy & Particulate Materials, 2000, pp. 5–147 to 5–157.

3

Developments in PM Engine Components

Cesar Molins, AMES Barcelona, Spain

Abstract

Automobile engines have undergone deep development in the past 20 years, seeking increased performance, reduced consumption, reduced environmental impact, reduced weight and volume, increased reliability and reduced cost.

PM has played an important role in the development of new engines either by improving existing components, by reducing their cost or by bringing components to newly developed systems which helped bring those goals to a reality.

Sources of improvement can be found in the raw materials, in the processes, in the machinery used to make the parts and in the tools and lubricants. The paper presents a number of examples of the innovations brought and used by PM and illustrates the way in which they have contributed to the mentioned goals.

3.1 SOME MAIN OBJECTIVES OF THE AUTOMOBILE MANUFACTURERS

Automobile manufacturers are in a constant race to outperform each other on one side and to comply with ever more restrictive regulations imposed by governments regarding safety and reduced environmental impact. All of this, added to an endless fight for cost reductions, along with the quest of special appeals related to styling or unique characteristics like off-road capabilities, ample space for passengers and/or baggage, etc.

Performance improvements are measured in a variety of ways:

- Acceleration
- Top speed

- Low consumption
- Stability
- Noise
- CO_2 and Nox emissions
- Braking performance
- Resistence to corrosion, etc.

All of these, added to the economic fight related with warranty (in terms of Kilometers or years or both) and sheer price tag. One obvious way to achieve several of these goals is to achieve higher output power with smaller displacement engines, as there is an immediate reduction of weight with the associated inertial inefficiencies and smaller displacements call also for smaller amounts of fuel just to maintain the engine running.

As of today, we can consider that all automobiles are powered by internal combustion engines, with hybrids or electrical cars representing yet an almost negligible proportion.

Internal combustion engines can be classified in two main categories: Otto cycle (gasoline fuel) and Diesel cycle (diesel oil fuel). Diesels have always offered smaller consumptions for the same power compared to gasoline engines by a factor of approximately 20%, due in a good part to the much higher compression ratio and to the absence of vacuum in the intake manifold. However, for many years diesel engines were slow in reaction compared to their gasoline counterparts, noisier and prone to exhaust significant soot, mostly during accelerations. These factors kept diesels to a reduced usage in many countries, except for lorries, buses and taxi cabs.

However, with the evolution of technology, diesel engines have increased significantly their power density, increased responsiveness, reduced noise and soot, thus surpassing otto engines' market share in car applications in regions such as Europe. In any case, otto engines have also significantly increased their power density in the last 20 yers. At this moment there is a practical tie in the power density, with a surprising slight advantage for the diesel engines, but both reaching levels at or above 100 HP/liter, as can be seen on Fig. 3.1.

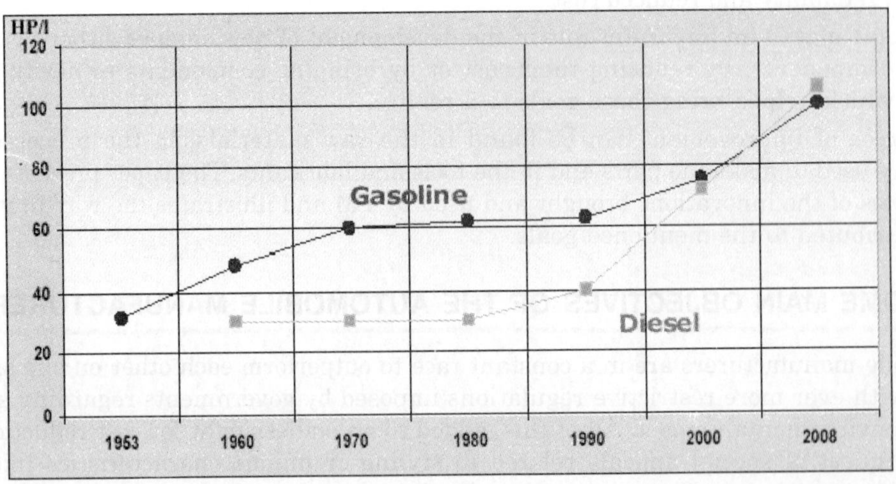

The technologies that have enabled this spectacular power surge are also responsible for the improvement of many of the other performance criteria. The means to achieve these goals can be sorted in to two main categories:

3.2 SOME MEANS TO ACHIEVE THOSE GOALS

On one side there are the improvements made on existing technologies and systems, while on the other would be totally new systems. In both cases, actions and evolution have been complemented by considerable cost reductions, which have enabled the inclusion of the new technologies or new systems in affordable cars. In all cases, PM played a significant role as an enabling technology.

The main players in the categories of improved existing technologies and in the new systems are the following:

Improvements on Existing Parts and Systems

- *Increased strength*: New alloys for metals, evolved plastics and composite materials have enabled miniaturization of parts to perform the existing functions.
- *Tightened tolerances*: The improvement of machinery and processes used to make the components have also allowed a reduction of tolerances which, in turn, were key in allowing longer lives for components, in increasing working pressures, reducing vibration and allowing miniaturization, thus reducing weight and cost of the components.
- *Increased wear resistence*: A clear factor influencing the life expectancy of components. With the market drive to extend warranties, some systems could only be considered after a certain life span could be assured.
- *Increased corrosion resistence*: In addition to influencing the sheer durability of some components exposed to harsh environments, the increase of corrosion resistence meant some times the availability of some safety-related systems, such as the tone wheels for ABS systems. In other cases and with different types of corrosion resistence, new materials enabled uses in hot and corrosive gases environments, such as in turbochargers or EGR systems.
- *Higher pressure hydraulics*: Normally enabled by reduced tolerancing and by the availability of more complex shapes thanks to improved compacting technologies, higher pressure hydraulics enabled smaller hydraulic systems (for example, for power steering), thus making them viable for smaller vehicles.
- *Reduced volume and weight starters*: Necessary for reduction of cost, weight and volume, they called for very efficient, small and inexpensive reduction gear, which was brought by improvements in geometries and materials.
- *Higher-speed magnetic actuators*: Enabled the use of much improved electronic systems.
- *Reduced-weight, improved reliability and faster electronics*: Modern cars and engines contain an unbelievable computing power which continuously optimize power, fuel and safety. The cost, weight and reliability of these electronics were key factors in enabling many of the engine-optimizing systems.

New Systems

- *VVT*: Variable Valve Timing. Allows the optimization of the opening and closing times of intake and exhaust valves, thus improving fuel economy and engine response at the different rpm regimes.
- *EGR*: Exhaust Gas Recirculation. System which significantly reduces emissions by improving burning conditions. Environmental impact and fuel consumption improvement as well.
- *Common rail*: System which has allowed a dramatic increase of top-rpm at diesel engines, thus increasing output power with the same engine and injection pressure, which was responsible for a significant improvement in the combustion conditions and thus, a reduction of soot and fuel consumption.
- *Variable flow oil pumps*: Oil pumps which reduce oil flow at high-rpm, when an oil flow proportional to rpm is excessive. The result is a reduction of consumed power and thus, of fuel.
- *Multi-jet injection systems*: Systems which allow a number of sequential injections within a single combustion phase in a cylinder. This results in much improved combustion, reduced fuel consumption and emissions and reduction of the once-traditional "clicketing" noise of diesel engines.
- *General Cost reductions*: In the past 15 years, industry has made major strides in reducing the cost of components. This enabled the use of some of the newly developed technologies to large-scale production models. The cost reductions were achieved through a variety of paths: More sophisticated machines for production, improved materials for the components and for the cutting tools, changes in design to allow simplification of the machining operations or, in some cases, allowing a change of technology to a lower-cost one, as was often the case in converting parts from machining to PM, which resulted in lighter, more efficient and cheaper parts. In many cases, the redesign of components led to consolidation of several parts into a reduced number, with frequent reductions of weight and cost, much of which was feasible thanks to the evolution of materials and the machinery used to manufacture the parts. In some cases, manufacturing processes were shortened thanks to the consolidation of parts. Here again, PM was a major contributor to this trend.

3.3 CONTRIBUTION OF PM TECHNOLOGY TO THE MARKET DEMANDS

A few key examples of the contributions of PM to the improvements of engines described above are the following:

3.3.1 Powder-forged Connecting Rods

Aimed at the beginning to a reduction of machining costs because of the much tighter tolerances achievable with the PM route compared to the traditional route, plus the feasibility of a fracture-splitting of the rod's main bearing head, it resulted in addition in an improvement of balancing in the engines, thanks to a smaller scatter of weights in the production lots.

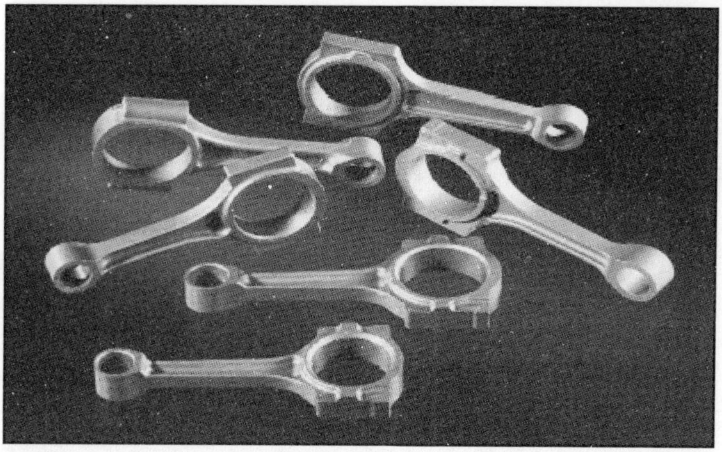

Courtesy of GKN

3.3.2 Main Bearing Caps

They were developed in PM also as a cost reduction for their improved geometry and reduced need for machining as compared to their cast counterparts. PM brought in addition a certain weight reduction, important for such massive components.

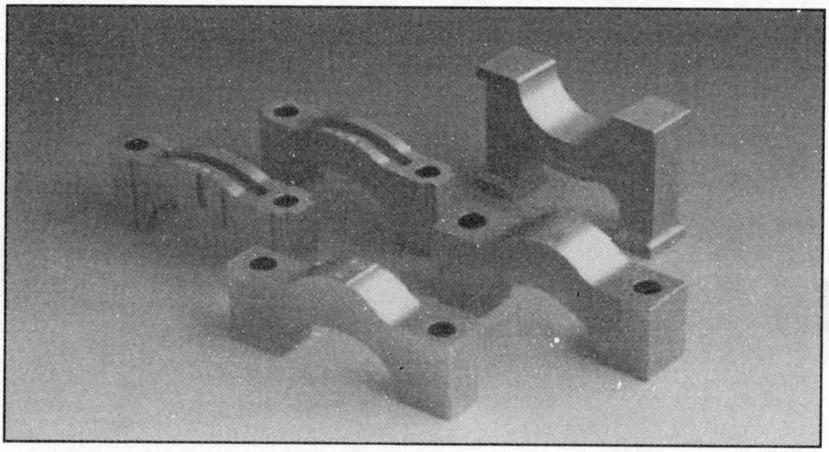

Courtesy of PMG

3.3.3 Aluminum Camshaft Bearing Caps

The evolution of PM in manufacturing Al alloys allowed the production of these parts, with significantly reduced weight compared to the previously wrought steel components and at reasonable prices thanks to the geometric capabilities of the PM technology.

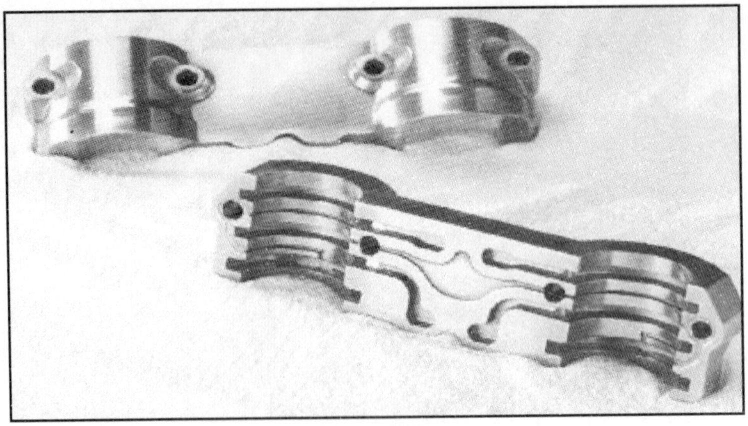

Courtesy of MPP

3.3.4 Gasoline Pump Elements

The substitution of carburetors by gasoline injection systems, which meant a significant reduction of fuel consumption and a reduced level of emissions required a total change in the pumps, which previously showed significant pressure oscillations. The new gerotor gasoline pumps, thanks to the PM technology, allowed much more constant fuel pressure, reduced weight, reduced noise and reduced cost. New materials as well as improved machining technology was necessary. Oter gasoline pumps, using plastic centrifugal impellers, also use sophisticated PM bushings to reduce friction. Much tighter tolerances were necessary to keep the required fuel pressure and flow.

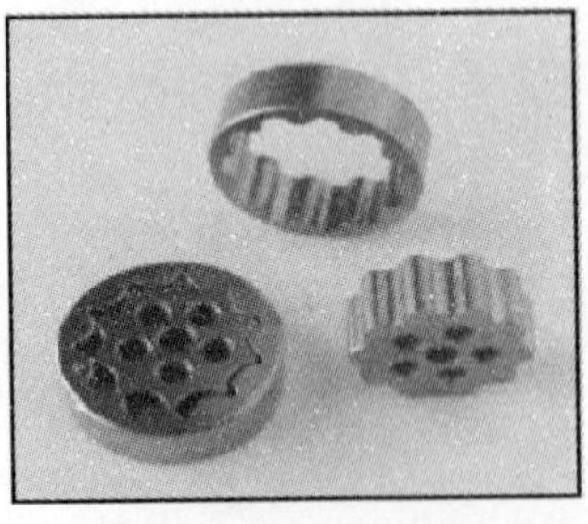

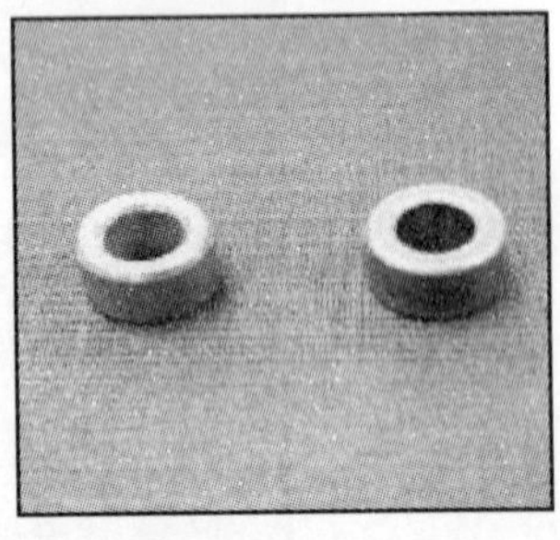

3.3.5 Variable Flow and Weight Reduced Oil Pumps

Traditional oil pumps create an oil flow which is proportional to the engine rotational speed. They have to be designed for the most unfavourable condition: running on hot oil at idle speed. Then, when the engine revs up, there is an excessive flow of oil which only causes increased power consumption, with no added function. New design, variable flow pumps reduce this effect and they are designed around the geometrical benefits of the PM process. Other pumps also use the PM design capabilities to reduce weight and material contents (thus also cost) and in some cases this design is also used to improve the filling of oil into the pumping cavity.

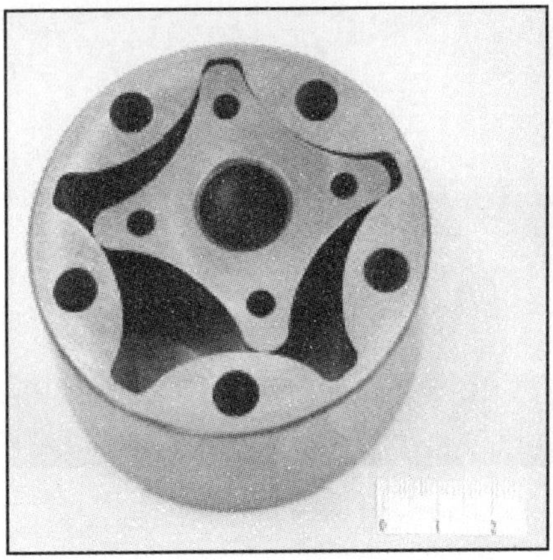

3.3.6. VVT: Variable Valve Timing

A novel technology was developed to optimize the opening and closing times for intake and exhaust valves, considering that the dynamics of gases in the combustion chamber are undoubtedly different at low rpm and at high rpm. The developed systems include a set of very complex and high precision mechanical parts which are located at the driving end of the camshafts and for which PM technology was instrumental. The VVT systems improve engine response and reduce fuel consumption and emissions.

Courtesy of SHW

Courtesy of MIBA

3.3.7 Composite Camshafts

Camshafts were traditionally forged from solid bars and heavily machined to the final shape. A new trend was to mount the cams and intermediate rings on a hollow tube. The result is a much lighter assembly and significant reduction in the machining costs. PM was the technology of choice to develop the cams and rings, which needed to meet stringent wear resistence and dimensional precision.

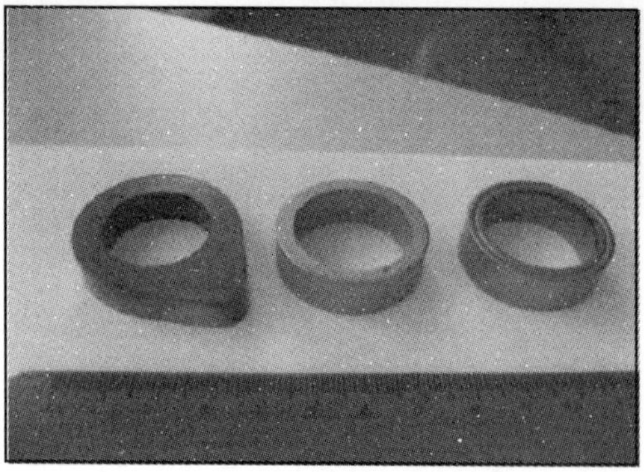

Courtesy of Thyssen

Courtesy of Thyssen

3.3.8 Lightweight Distribution Pulleys

Distribution pulleys have been made out of PM since many years ago, but the development of more sophisticated presses and tool technology has enabled the geometrical evolution of pulleys to optimize space in the engine and to make them lighter and lighter, with the corresponding savings in cost of materials, weight and inertial masses

3.3.9 Common Rail Diesel Injection Systems

Common Rail fuel injection meant a major revolution in the performance of diesel engines, bringing them to much higher rpm, increasing significantly injection pressure from the old 200 bar to 1600 and now approaching 2000 bar, which greatly improves pulverization of fuel and thus quality of combustion, reducing fuel consumption and emissions. The injection pumps require a first-stage fuel pump which is normally made with PM components demanding tight tolerances and excellent wear resistence.

3.3.10 EGR: Exhaust Gas Recirculation Systems

EGR is a system which recirculates part of the exhaust gases back to the combustion chamber in order to reduce the amount of NOx in the exhaust. It needs valves and actuators working at high temperatures with combustion gases, which are rather aggresive. PM has also been instrumental for some of the parts in these systems, such as sliding bushings or valve bodies. Some of these parts are stainless steels or special bronzes.

3.3.11 Stainless Steel Exhaust Flanges

The changes in combustion conditions and fuel compositions have caused an increase of the temperature of exhaust gases. Also, the stricter demand of durability from the market have created a need for some of the exhaust flanges to be manufactured in stainless steels. PM was able to respond to this new market demand and PM stainless steels of today are significantly more resistant to corrosion than they were in the past. The new products, even though more expensive than normal steel end up in savings due to the increased durability.

3.3.12 Gasoline Injection System

The last 20 years have seen an almost absolute transition from the carburetor system to the gasoline injection system, due to the reduction of fuel consumption and the associated reduction of emissions enabled by the gasoline injection. Soft magnetic PM materials have been widely used in the injectors, where fast response and low losses were necessary, together with complex geometries and sometimes special characteristics, like elongations above 10% in order to allow crimpings.

3.3.13 Increased Pressure Power Steering

With the evolution of compacting presses and improved tolerances in toolmaking, power steering parts became more complex, more dense and tighter in their tolerances, which enabled the increase of hydraulic pressure in the system and thus, the reduction of hydraulic element dimensions. Power steering has long been a clear demonstrator of PM's geometric capabilities and evolution.

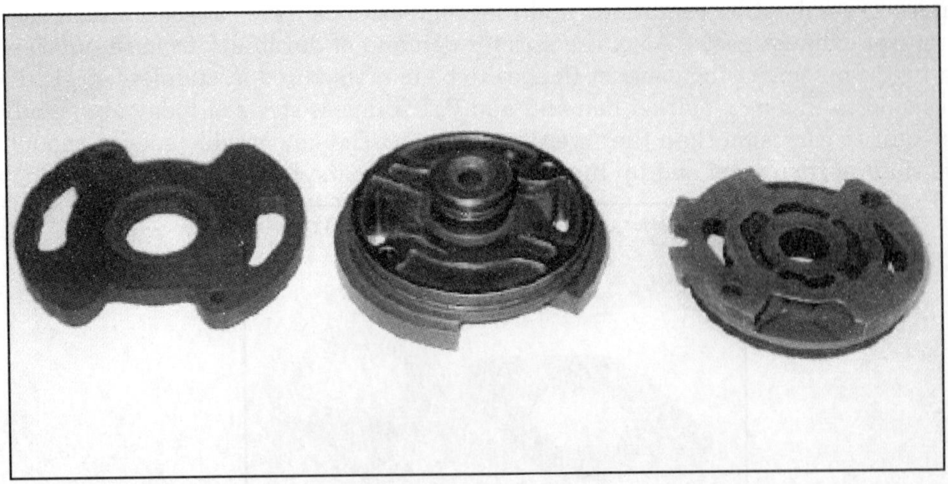

3.4 EVOLUTION OF NECESSARY TECHNOLOGIES IN PM

The achievement of the examples mentioned above, which enabled improvements in the engines in terms of performance, cost, reliability, consumption and eco-friendliness needed the development of a number of materials and technologies in the PM raw materials, machinery and ancillary products.

Some of the main developments are the following:

3.4.1 Sinter Hardening Steels

These steels (which required the development of newly designed furnaces) allow in some cases the avoidance of a hardening heat treatment, thus reducing costs in the manufacturing process and sometimes improving dimensional stability. The down sides are a lower hardness achieved and the frequent through hardening of the parts.

3.4.2 Low Alloy, Low Cost, High-strength Alloys

The dramatic excursions of some metals in recent years plus the fears that Nickel could be declared a carcinogenic agent in spite of its very wide use in everyday's utensils triggered the development of alternative composition steels, such as Cromium steels. They present, however, some difficulties for sintering and the ranges of physical properties achieved fall short of those available with the well-known Cu-Ni-Mo steels.

3.4.3 Improved Soft-Magnetic Materials and their Processing

The ever incresing presence of electronics in the automobiles and in the engine's control call for more soft-magnetic actuators and sensors. High-frequencies are becoming more common and new materials have been and are still being developed to respond to this increasing technological demand. Some have good responses at medium frequencies, but show low mechanical strength.

3.4.4 Sinter-Hardening Furnaces

Were developed to enable the processing of the steels mentioned in 3.4.1. They have been perfected through several years of use and design improvement.

3.4.5 Sinter Forging

Was a response to the continued demand for higher mechanical strength and improved fatigue resistence. Sinter forged parts show some advantages versus their wrought counterparts, but sometimes they have trouble competing with prices, due to the cost of the PM raw materials.

3.4.6 Press Hardening

This is a technology which was already well known for heat treatment of slender and precise heat treated wrought parts, but of little application in PM. It now appears to be gaining new applications.

3.4.7 Warm Compaction

This is aprocess that was developed to avoid some coining a operations or sometimes double-press, double-sinter processes for parts which could do with an intermediate density. It took years to perfect the raw materials and the machinery to do the warming up, but it is considered today a well-established technology.

3.4.8 Improved Compacting Lubricants

There are today some lubricants which allow reaching densities above 7.4 g/cc in steels with a single press stroke. This is a definite step forward in the race for higher performance steels and reduced process costs.

3.4.9 Improved Coining Lubricants

There are also today coining lubricants which use no volatile components, thus having a reduced environmental impact and often reducing the oily dirt around the coining press, which eventually results in reduced usage of lubricant and reduced cost.

3.4.10 Improved Precision Presses and CNC Controls

Compacting presses have experienced dramatic improvements in recent years. There are presses which can ensure punch positionings down to a few microns, thanks to the improved encoders and electronics.

3.4.11 Increased Number of CNC Plattens in Presses

There are presses with up to eight lower CNC punches and 3 upper punches, actuated by hydraulic means at speeds that were previously only reachable with mechanical presses. Electrical power consumptions have also been dramatically reduced.

3.4.12 Improved Tool Geometry and Tolerancing

The improvements in presses mentioned above, together with the also dramatic improvements in tool-making machine tools have enabled unheard of geometries and tolerances down to just some microns, thus improving the compact's precision and often the tool's lives.

3.4.13 Surface Densification

This is a technology which is still in full evolution. Just a few PM parts makers in the world can claim to have surface densification in industrial operation. Tests and studies made on the subject seem to prove higher fatigue resistence, thus opening possibilities in applications which were previously unreachable for PM production.

3.4.14 Improved Stainless Steels Processing

Both the raw materials and the furnaces have improved to the point that PM stainless steels show today corrosion resistences several times higher than those achievable a few years ago. There are PM stainless steels that can withstand close to 1000 hours in salt spray chamber, and even above this figure in some laboratory environments

3.4.15 New Sealing and Plating Technologies

Some times surface porosity of a PM part may represent a problem. Sealing products and processes have also improved, giving better sealings and cleanliness of the parts.

3.4.16 More Sophisticated Machining Machines

This is not exclusive to the PM technology, but is also used for secondary operations on PM parts. Faster, more precise and more reliable machine tools enable today some tolerances that were unfeasible a few years ago.

3.4.17 Improved Cleaning Machines and Cleanliness Control Systems

Open porosity in PM parts is sometimes also a problem in terms of contamination for some clean-environment applications. Better cleaning machines and fluids, along with better cleanliness evaluation methods allow the use of PM parts in applications which were hardly reachable before.

3.4.18 Novel Non-destructive Testing

PM parts may present problems of hidden cracks or internal inhomogeneous porosities, which need to be 100% tested in some cases, obviously without destroying the part being tested. Sound testing and ultrasonic resonance are two technologies which have experienced important evolution for testing of these defects.

3.4.19 Novel On-line and Off-line Inspection Systems

Customers are increasing their demands on fully defect-free production. Just a few years ago, a production with 30 ppm deffect rate was considered remarkable. Today, in some applications customers will seriously object to 1 ppm defect. This is hard to ensure and novel inspection techniques have appeared in the market, such as computerized vision systems or laser interferometry flatness mapping.

3.5 CONCLUSIONS

- Car engines have undergone major improvements in the recent past.
- The results are more powerful engines, higher fuel efficiency, lower environmental impact, reduced weights and lower costs.
- Power Metallurgy has played a significant role in the evolution of engines.
- New materials, new processes, new geometries and tighter tolerances have been brought to the market.
- To achieve this, new materials have been developed, new machines with dramatically improved control and positioning systems and new measuring technologies have been brought to the market.
- All of this has been possible through the cooperation of industry and research institutions encompassing a wide variety of knowledge fields.
- PM continues to be a very trustworthy and economically attractive technology for future developments.PM is in addition an environmentally and energy-wise attractive technology.
- PM is in addition an environmentally and energy-wise attractive technology.

❍❍

4

PM Materials for Gear Applications

Senad Dizdar and Pernilla Johansson, Höganäs AB, Swedan

Abstract

This paper reviews recent achievements in gear tooth bending strength and rolling contact fatigue resistance of powder metal materials in comparison to wrought steel materials for automotive applications. Prototype powder metal gears and rolling contact fatigue (RCF) rollers were pressed from commercially available low-alloyed iron powder mixes, sintered, surface densified, case hardened and optionally hard finished using commercially available equipment for all manufacturing steps. The reference gears and rollers were machined from common wrought steels. The results evaluated on the prototypes show that powder metal materials meet wrought machined materials' gear tooth bending strength and RCF-resistance.

4.1 INTRODUCTION

Developments in the automotive industry toward relatively inexpensive, high performance vehicles with low fuel consumption and low environmental impact put high demands on materials for automotive gear applications. Such developments require compact size but high strength automotive gearboxes. The gears that meet automotive industry demands are made of low alloyed steels for case hardening such as AISI 5115 (DIN EN 16MnCr5) or 8620 (21NiCrMo2). The gears are manufactured by a process route consisting of three principal operation steps: soft machining, case hardening and hard finishing. These gears achieve higher than 1000 MPa in gear tooth bending strength σ_{FE} and 1500 MPa in gear pitting resistance σ_{Hlim} per ISO 6336.

Conversely, powder metal materials have, until one or two decades ago, been associated with low cost and low performance gear applications, such as pumps, hobby and household applications. However, the introduction of high density technologies improved gear density levels and by this their mechanical strength. Warm compaction [1] improved density levels of parts such as power tool's gears up to 7.2–7.3 g/cm^3, and offered a lower cost alternative to double press double sinter (DPDS) route. High velocity compaction [2] offered the possibility to cost effectively single press large single level powder metal parts such as parking gears up to 7.2–7.3 g/cm^3. Finally, gear surface densification techniques [3] open possibilities to fully densify the tooth surface to a depth that Hertzian contact stresses in the gear flank and bending stress gradients in the gear tooth root are within it. This paper has been aimed to review recent achievements in high performance metal powder gears manufactured by powder metallurgy routes consisting of pressing, sintering, surface densification and case hardening.

4.2 EXPERIMENT AND MATERIAL

Performance levels that powder metal gears achieve were evaluated on simple spur gears from an automotive application and rollers from RCF test by ZF Friedrichshafen. Figure 4.1 shows the prototype gear, basic facts about the gear tooth root bending testing and a brief schedule of gear bending stress calculations according to ISO 6336. Figure 4.2, illustrates the ZF-RCF test and lists some facts about the RCF-testing.

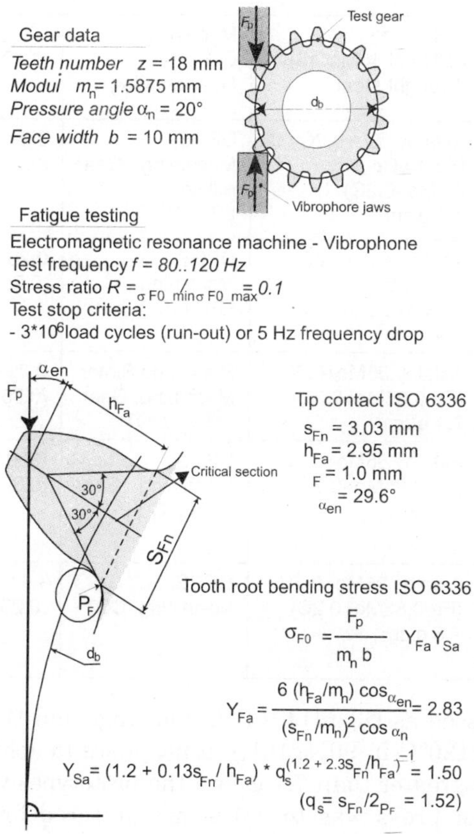

Fig. 4.1: The prototype gear, its gear tooth root bending testing and a brief schedule of gear tooth bending stress calculations

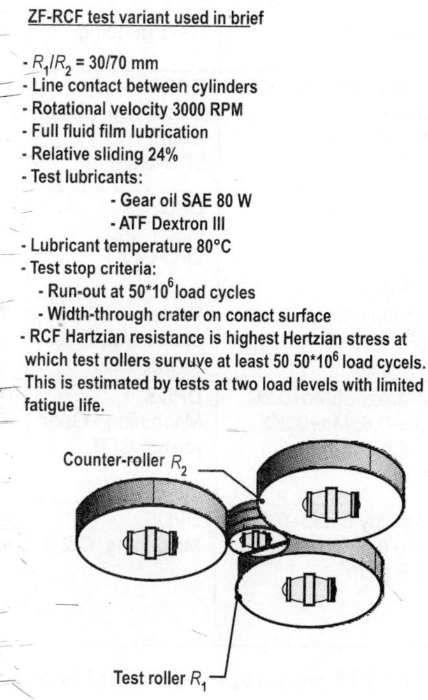

Fig. 4.2: ZF-RCF test in brief

The PM materials used in this study were the atomized prealloyed powder grades Astaloy 85Mo and Astaloy CrL. Astaloy 85Mo is prealloyed with 0.85% Mo while Astaloy CrL is prealloyed with 1.5% Cr and 0.2% Mo.

The manufacturing process route for each prototype gear and roller are described in Table 4.1 and Table 4.2. The reference gears and rollers were machined from wrought round bars and case hardened at a heat treat vendor. The case hardening (CQT) was performed according to common practice for components made of low alloyed steels aimed for case hardening *i.e.*, they were gas carburized for 30 and 150 minutes at 920°C, quenched in 60°C oil bath and air tempered at 200°C for 60 minutes. The common DIN 550 HV casehardened depth ($E_{ht5550HV}$) was used. The reference gears were finish ground to achieve gear quality DIN 7.

Table 4.1

Material, manufacturing routes and achieved surface densification and hardness of the test gears.

	Material and base (core) density	Manufacturing route	$SDD_{0.98}$/ Eht_{550HV} (mm/mm)
1	AISI 5115 (DIN EN 16MnCr5) Wrought steel	Machining, CQT, Gear grinding	- /0.19
2	Astaloy CrL+0.2C (Fe-1.5Cr-0.2Mo+0.2C) 7.1 g/cm³	DPDS, Machining, Gear rolling LPC-GQ-T	0.20 /0.25
3	Astaloy CrL+0.2C (Fe-1.5Cr-0.2Mo+0.2C) 7.4 g/cm³	DPDS, Machining, LPC-GQ-T	0.20 /0.25
4	Astaloy 85Mo+0.2C (Fe-0.85Mo+0.2C) 7.1 g/cm³	Press and Sinter Machining, Gear rolling, CQT	0.20 /0.30
5	Astaloy 85Mo+0.2C (Fe-0.85Mo+0.2C) 7.4 g/cm³	DPDS, Machining, Gear rolling, CQT	0.20 /0.32
6	Astaloy 85Mo+0.2C (Fe-0.85Mo+0.2C) 7.6 g/cm³	DPDS, Machining, CQT	- /0.25

Table 4.2

Material, manufacturing routes and achieved surface densification and hardness of the test gears.

	Material and base (core) density	Manufacturing route	$SDD_{0.98}$/ Eht_{550HV} (mm/mm)
1	AISI 5115 (DIN EN 16MnCr5) Wrought steel	Machining, CQT, Gear grinding	- /0.19
2	Astaloy CrL+0.2C (Fe-1.5Cr-0.2Mo+0.2C) 7.1 g/cm³	DPDS, Machining, Gear rolling LPC-GQ-T	0.20 /0.25
3	Astaloy CrL+0.2C (Fe-1.5Cr-0.2Mo+0.2C) 7.4 g/cm³	DPDS, Machining, LPC-GQ-T	0.20 /0.25
4	Astaloy 85Mo+0.2C (Fe-0.85Mo+0.2C) 7.1 g/cm³	Press and Sinter Machining, Gear rolling, CQT	0.20 /0.30
5	Astaloy 85Mo+0.2C (Fe-0.85Mo+0.2C) 7.4 g/cm³	DPDS, Machining, Gear rolling, CQT	0.20 /0.32
6	Astaloy 85Mo+0.2C (Fe-0.85Mo+0.2C) 7.6 g/cm³	DPDS, Machining, CQT	- /0.25

The PM prototype gears and rollers were pressed as D34/d14 × h10 mm rings and D40 × 28 mm cylinders and sintered for 30 minutes at 1120°C in $90N_2/10H_2$ atmosphere to achieve the nominal core density. To achieve core densities greater than 7.1 g/cm³, the prototypes were double pressed double sintered (DPDS). The first press was to 7.0 g/cm³, sintered for 30 minutes at 800°C in $90N_2/10H_2$ atmosphere, then re-pressed to 7.4 and 7.6 g/cm³ densities and then sintered for 30 minutes at 1120°C in $90N_2/10H_2$.

The rings/cylinders were then machined to gear and roller blanks with rolling–ready–geometry. This means that the gear blanks were hobbed using a gear hob with intentionally modified profile to a geometry that includes allowance for flank surface material to be displaced by the gear rolling die. A typical oversize for tested gears was 0.3 mm in the over ball measure (OBD).

The gear blanks were then rolled using a commercial CNC radial (transverse) rolling machine (Fig. 4.3) and achieved 0.3 mm surface densified layer depth – the depth in millimeters at which pores occupy 98% of the relative full density ($SDD_{0.98}$) (Fig. 4.4). The PM gear rolling process is analogous to soft finishing of wrought gear [4], a process used by some auto manufacturers. The PM roller blanks had an oversize diameter of about 0.5 mm for rolling allowance, and were rolled using the same machine as for the PM gears. Reference [5] includes an overview of PM rolling technology. All PM prototypes except those made of Astaloy CrL were case hardened at the same heat treat vendor using the same furnace and the same parameters in order to achieve compressive residual stresses by creating a hard martensitic surface with a soft core.

The gears made from Astaloy CrL were low pressure carburized, gas quenched and tempered (LPC–GQ–T) by a heat treat vendor. Carburizing was done at 960°C using acetylene gas with flow 700 nl/h at pressure of 3 mbar under boost cycles of 4, 2 and 2 minutes separated by diffusion cycles of respective 6, 11 and 40 minutes. The gears were then gas quenched with nitrogen under 10 bar pressure. However, this process was not optimized for the gears.

Fig. 4.3: A photograph of a P/M gear rolling in a CNC radial rolling machine.

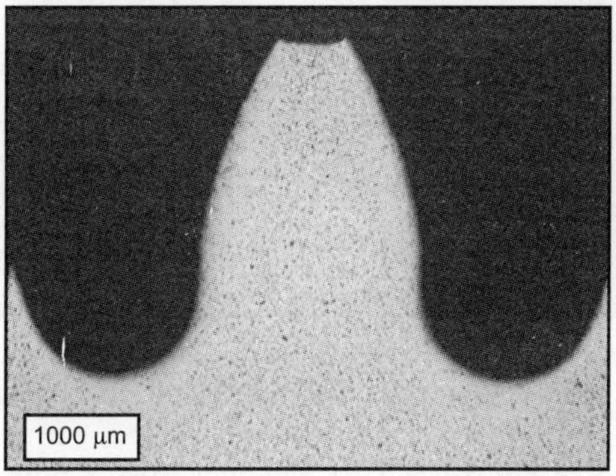

Fig. 4.4: Pore structure of a surface densified gear tooth made of Astaloy CrL + 0.2C, with 7.4 g/cm^3 core density and densified depth of 0.3 mm.

Microstructure in case hardened Astaloy 85Mo materials (Fig. 4.5) consists of plate martensite with high carbon content at the surface and lath martensite in the core with low carbon content. Between the surface and the core there will be a carbon content gradient where the martensite gradually will have lower carbon content. Case hardened Astaloy CrL material (Fig. 4.6) is characterized by plate martensite with high carbon content at the surface and deeper in the material the carbon content will decrease and the plate martensite will be mixed with lath martensite. Eventually the lath martensite will be mixed with more and more bainite so in the core the microstructure will be bainite mixed with lath martensite.

Microhardness profiles of tested gears are shown in Fig. 4.7. Comparing to the reference AISI 5115 gears, surface densified Astaloy CrL gears have higher surface hardness but lower core hardness while Astaloy 85Mo gears have similar surface hardness but lower core hardness. The reasons for such microhardness picture are likely core porosity and alloying content.

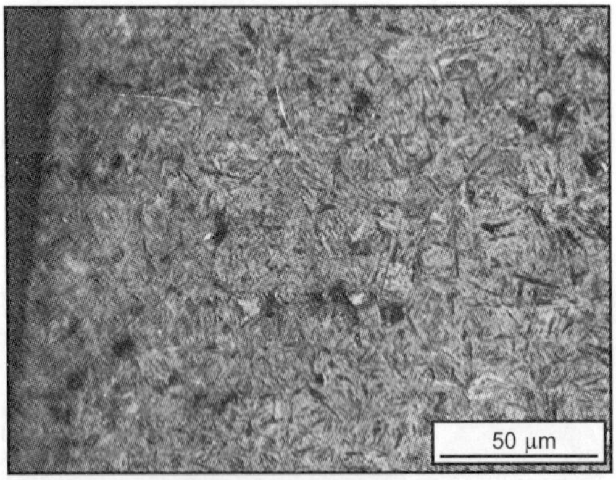

Fig. 4.5: Etched structure of the tooth flank surface of an Astaloy 85Mo + 0.2C, surface densified case hardened gear tooth

Fig. 4.6: Etched structure of the tooth flank surface of an Astaloy CrL + 0.2C, surface densified case hardened gear tooth

Microhardness profiles of the tested RCF-rollers are shown in Fig. 4.8. All the rollers, even those made of Astaloy CrL, were gas carburized, oil quenched and air tempered.

Microhardness profile of the surface densified rollers made of Astaloy CrL + 0.2C, 7.1 and 7.6 g/cm^3 density, are close to the microhardness profile of the reference AISI 8620 rollers. Microhardness profile of Astaloy 85Mo + 0.3C, 7.0 g/cm^3 density rollers showed a considerable difference from the others. The surface densified gears achieved a very high surface hardness while the non-surface densified gears achieved a micro-hardness profile with a surface hardness of slightly over 600 HV0.1. A general comment for the roller's case hardening is that surface densified powder metal rollers behave the same as full dense wrought material rollers, while non-surface densified rollers powder metal with density level of approximately 7.0 g/cm^3, and interconnected pores, more difficult to case hardening results of the gear tooth bending testing are shown in Fig. 4.9 and Fig. 4.10. Case hardened surface densified gears made of Astaloy 85Mo + 0.2% C show a high gear tooth bending performance. Those with core density of 7.4 g/cm^3 have even higher performance than case hardened wrought machined AISI 5115 gears. The Astaloy CrL + 0.2% C gears have slightly lower performance but case hardening for this material was not optimized at the time of publication. The effect of surface densification on the gear tooth bending performance appears to be a gain of 100 MPa in gear tooth bending strength when core density increases from 7.1 to 7.4 g/cm^3. However, absence surface densification lowers the gear tooth bending strength down to 850 MPa level despite a high core density of 7.6 g/cm^3.

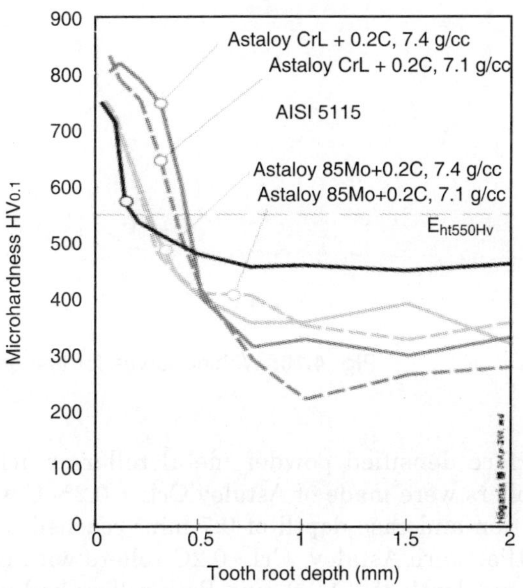

Fig. 4.7: Microhardness of tested gears

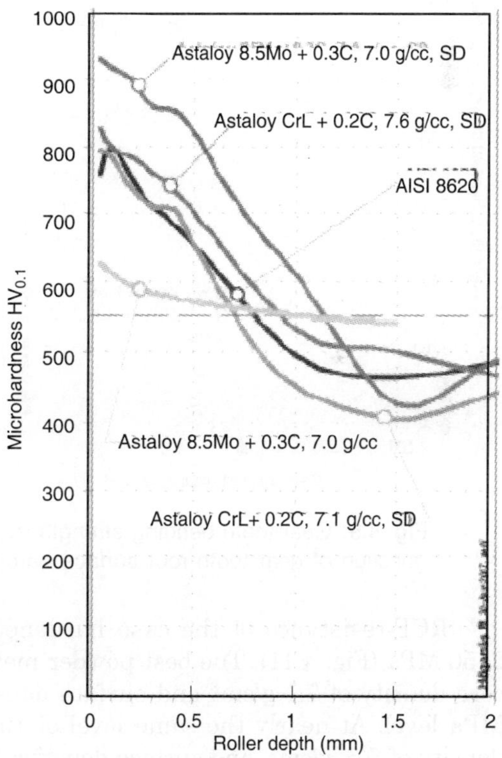

Fig. 4.8: Microhardness profile of tested RCF-rollers

When comparing gear tooth bending strength of different gear sizes *i.e.*, modules, the size effect has to be considered. ISO 6336 includes gear tooth bending strength levels evaluated on larger wrought gears with module 3–5 mm depending on core hardness level and Ni content according to ISO 6336. The test gears in this investigation had a relatively small module of 1.5875 mm (DP 16). Their gear tooth bending strength exceeded ISO 6336 levels for 3–5 mm module gears. The same phenomenon was previously reported by *e.g.*, Jeong in 1992 [6], who found a difference in casehardened gear tooth bending strength of 22% between modules of 1.5 and 5 mm. Dividing 1350 MPa in gear tooth bending strength of casehardened AISI 1515 gears with the core hardness of 440 HV0.1 by 1100 MPa for casehardened wrought gears with core hardness of 40 HRC (ca. 400 HV), yields a difference of 23%. This is additional evidence of validity of the results shown here.

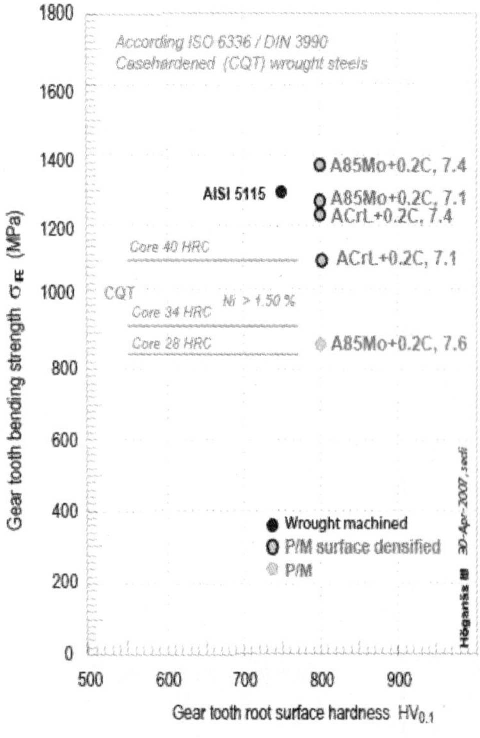

Fig. 4.9: Gear tooth bending strength σ_{FE} as a function of gear tooth root surface hardness.

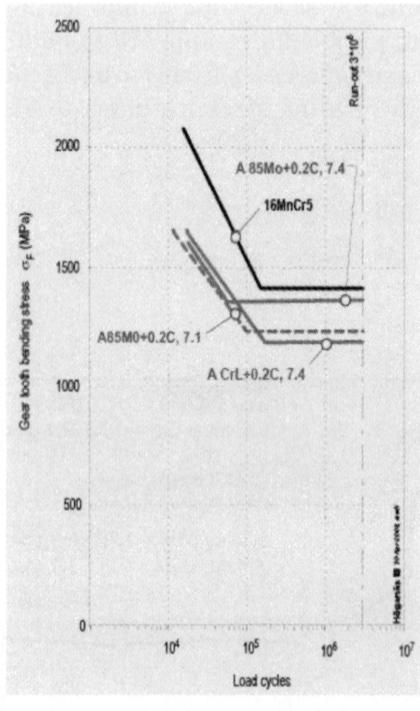

Fig. 4.10: Wöhler curves for test gears

RCF-resistance of the case hardened surface densified powder metal rollers reached 2150 MPa (Fig. 4.11). The best powder metal rollers were made of Astaloy CrL + 0.2% C with core density of 7.1 g/cm^3 and surface densification and case depth of 0.7 mm reached 2100 MPa level. At nearly the same level of 1950 MPa, were Astaloy CrL+0.2C rollers with core density of 7.6 g/cm^3, and surface densification/case depth of 1.4/0.9 mm. Both rollers had very similar profiles of residual compressive stresses down to 0.5 mm depth. The reasons for this 150 MPa difference in the RCF-resistance can be found in other aspects of case hardening quality, uniformity of surface densified depth, ratio of case hardened and surface densified depth and core density/hardness differences.

Astaloy 85Mo+0.3C rollers with density of 7.0 g/cm³, case hardened but not surface densified reached only about 1000 MPa. Similarly manufactured rollers surface densified, reached 1800 MPa illustrating how surface densification increases the RCF-resistance. Benefits of the surface densification are also illustrated in the slope of the Wöhler curves in the limited RFC-resistance region for 50% survival life according to log normal probability (Fig. 4.12). Surface densified Astaloy CrL+0.2C, 7.1 g/cm³ roller's curve overlie the reference AISI 8620 curve while the non-surface densified rollers made of Astaloy 85Mo+0.3C with density of 7.0 g/cm³ are significantly lower.

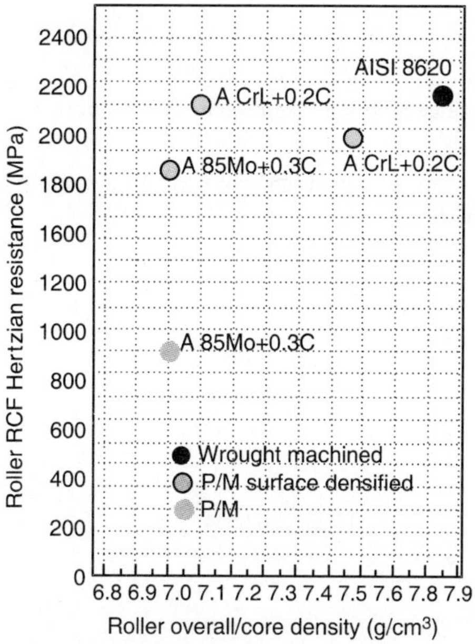

Fig. 4.11: Roller RCF resistance as a in function of surface hardness

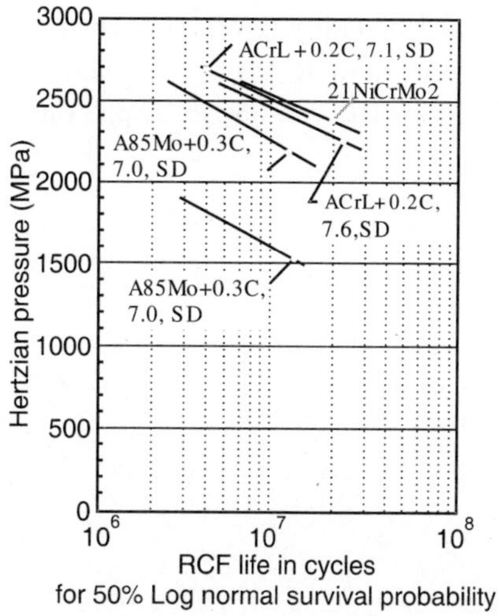

Fig. 4.12: Slope of the Wöhler curve the limited RFC-resistance region.

4.3 CONCLUSIONS

Powder metal materials for gear applications have been reviewed based on results of investigations into gear tooth bending strength and rolling contact fatigue resistance. Following conclusions were reached:

(1) Surface–densified case–hardened powder metal gears reached gear tooth bending strength of reference case–hardened machined wrought steel.

(2) Surface densified case hardened powder metal rollers reached RCF-resistance of case hardened wrought machined rollers.

(3) Surface densification plays the main role in reaching high performance of powder metal components.

(4) Chromium low alloyed powder metal materials show a promising potential for a high performance/cost ratio after optimization of case hardening.

REFERENCES

1. U. Engström, New high performance PM applications by warm compaction of Densmix powders, EURO PM2000, Munich 2000.
2. P. Skoglund, High density PM parts by high velocity compaction, Powder metall. 44/3 (2001), pp. 199–202.
3. P.K. Jones, K. Buckley-Golder, H, David, D. Sarafinchan, R. Shivanath, L. Yao, Fatigue properties of high density powder metal alloy steels for high performance power train applications, Proc. 1998 PM World Congress, pp. 155–166.
4. J.P. Dugas, Gear finishing by shaving, rolling and honing, Dudley's Gear Handbook, McGraw-Hill, 1991, pp. 18.1–18.33.
5. G.A. Kotthoff, Neue Verfahren zur Tragfähigkeitssteigerung von gesinterten Zahnrädern, Ph.D. Thesis, Technical University (RWTH) Aachen, (2003).
6. B. Jeong, M. Kato, K. Imoue, N. Takatsu: The bending strength of carburized fine module gear teeth, JSME Series III, Vol. 35 No. 1, pp. 136–141 (1992).

❍❍

5

An Approach to Reduce Costs of Heat Treated Powder Metallurgy Parts

Kalathur Narasimhan, Michael Marucci, Sunil Patel[1], and Narsi Chandrachud[2]
1. Hoeganaes Corporation., U.S.A. 2. GKN Sinter Metals, Pune, India

Abstract

Powder metallurgical parts are often heat treated to maximize performance in demanding automotive and non-automotive applications. The process 1. consists of compaction of powder, sintering and cooling the part and 2. reheating to austenitizing temperature and quenching followed by tempering (Q&T). Incorporating a rapid cool in process step1 can eliminate the process step two. The material selection for each of these processes is different. In a market place where prices of raw materials are fluctuating judicious selection of a process is required. Further, post-processing requirements should be considered in the selection process. In this presentation a few case studies on molybdenum bearing PM steels will be presented.

5.1 INTRODUCTION

Powder metallurgy is a cost effective process and can replace stampings, forgings, castings and machined parts. Today average powder content in automobiles is about 21 kilos in North America and about 9 kilos rest of the world (1–3). The content is smaller in India as the powder metallurgy process advantage is not well understood by the end users.

Cost is an important factor in the application of powder metal and efforts continue in reducing the cost of processing. Sinter hardening is an important factor in reducing the costs as it eliminates the need for heat treatment for powder metal parts.

However sinter hardening requires a rapid cooling section at the end of the sintering cycle to rapidly cool the parts and achieve about 90% martensite . Although cooling zones can be added to existing furnaces there is certain amount of reluctance from the parts makers. Additionally there is a misunderstanding on the part tolerances achievable through sinter hardening compared to quench and tempered parts. Post processing of sinterhardened parts is difficult if parts need additional machining. Heat treated (Q and T) parts can be machined before the heat treatment when the parts are in softer condition. Green machining is an option available for sinterhardenable parts. Alternately presintering the parts to remove the lubricant at temperatures as low as 600°C increases the strength of powder metal parts, which could facilitates the ease of machining. In this article we address sinterhardening steels and also heat treated PM steels.

5.2 SINTER-HARDENING ALLOYS

Specific P/M materials have been developed for the sinter-hardening process. These steel powders generally have a higher hardenability (the relative depth which a steel is able to completely transform to martensite) than conventional P/M steels [4]. The alloy can be introduced in several different ways (5) including a conventional admixture where elemental additions are made to pure iron powder. The alloy can also be diffusion alloyed where the elemental additions are annealed together with the iron powder. This bonds the alloy to the iron. The alloy can be prealloyed where the alloying elements are added directly to the molten steel prior to atomization. The final and most common method of alloying used for sinter-hardening is a hybrid alloy where prealloyed steel is admixed with elemental additions to produce the best compromise between hardenability and processability. This method also has the added benefit of avoiding the use of admixed Ni powder. Ni can be added directly to the base iron powder, this produces a more uniform and hardenable microstructure and limits the dusting of fine Ni powder into the air. Each alloying technique changes the compressibility of the material. This is displayed in Fig. 5.1(a). Note the increase in compressibility as the level of prealloying decreases.

Traditionally sinter-hardening alloys consisted of standard low alloy iron powders mixed with high concentrations of Cu, Ni, and carbon. The standard MPIF grades FLC-4608 and FLNC-4408 (Outlined in Table 5.1) are good examples of traditional sinter-hardening grades [7]. While these alloys offer excellent hardenability they also have several drawbacks. For instance, FLC-4608 uses a base alloy originally designed for powder forging. This alloy is optimized for fully dense hardenability, but has reduced compressibility due to the extensive use of prealloyed Ni. FLNC-4408 has better compressibility but is susceptible to dimensional variation due to the large amount of admixed elements. It also possesses a heterogeneous microstructure containing non-martensitic soft areas. Figure 5.1(b) shows a comparison hardness and absolute dimensional change for these alloys at a given set of processing conditions. The FLNC-4408 has the largest growth with the lowest apparent hardness.

Table 5.1: A selection of SINTER-HARDENING powder compositions

Designation	Base Iron	Base Iron Composition (w/o)	Admixed Cu (w/o)	Admixed Ni (w/o)	Admixed Graphite (w/o)	Admixed lube (EBS) (w/o)
FLNC-4408	Ancorsteel 85HP	Fe-0.85 Mo	2.0	2.0	0.9	0.75
FLC-4608	Ancorsteel 4600V	Fe-0.55 Mo-1.8 Ni-0.15 Mn	2.0	–	0.9	0.75
1.4Ni-1.25 Mo-0.4 Mn-2 Cu–0.8C	Ancorsteel 737SH	Fe-1.25 Mo-1.4 Ni–0.42 Mn	2.0	–	0.9	0.75

Specialized sinter-hardening base iron materials have been developed that offer a compromise between hardenability and processability. For instance, Ancorsteel 737SH(8), a Fe-Mo-Ni-Mn alloy offers better higher hardenability and compressibility than more traditional sinter-hardening alloys (8). This can be seen in Fig. 5.1(a).

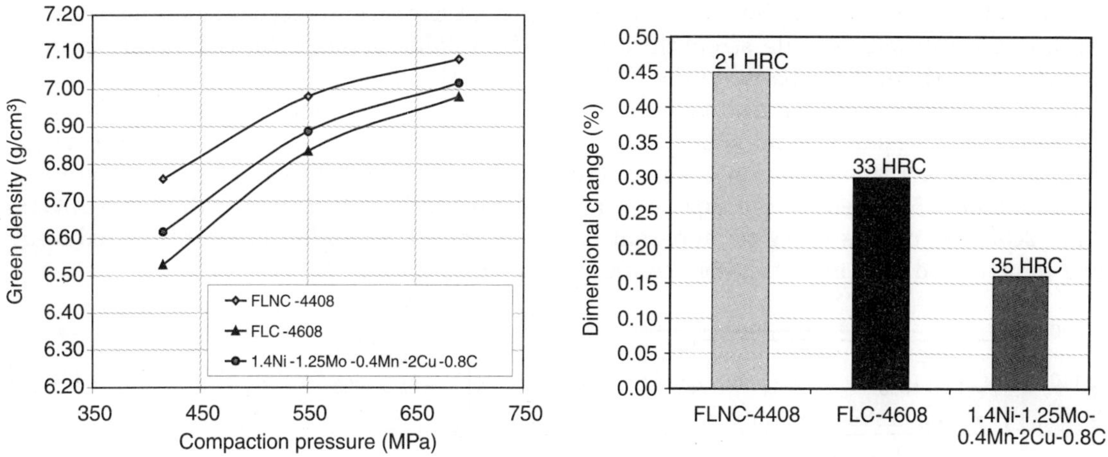

(a) (b)

Fig. 5.1: (a) Compressibility of select sinter-hardening materials, and (b) Comparison of dimensional change and hardness properties of select sinter-hardening materials. (Sintered at 1120°C in 90v/oN_2-10v/oH_2, Tempered at 200°C for 1 h)

Figure 5.2 shows comparative microstructures of several sinter-hardening steels. Note how the microstructure changes as the type of alloy changes. Figure 5.2(a) shows how the use of admixed Ni leads to non-martensitic Ni rich areas (light etching – white areas). These are soft and lower the overall hardness of the material. Also, the admixed alloy has not completely diffused into the steel leading to areas of lower hardenability, resulting in bainite/pearlite regions (dark etching – blue areas). Figure 5.2(b) shows that prealloying the Ni eliminates the Ni rich areas, however the hardenability is not sufficient to complete the martensitic transformation The micrograph in Fig. 5.2(c) shows that the higher hardenability steel admixed with Cu results in

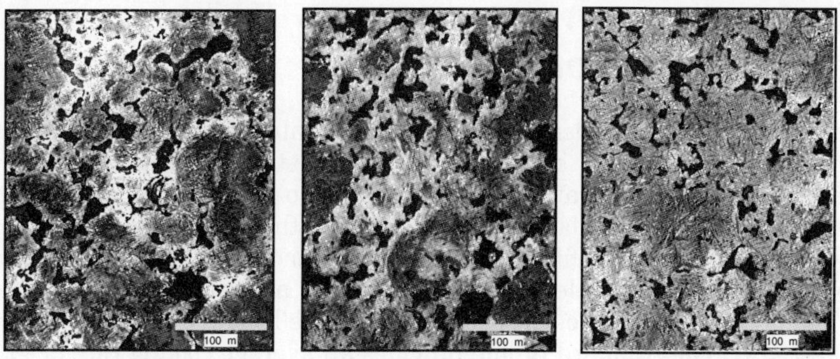

(a) (b) (c)

Fig. 5.2: Etched optical photomicrographs of (a) FLNC-4408, (b) FLC-4608, and (c) 1.4Ni-1.25Mo-0.4Mn-2Cu-0.8C, (Sintered at 1120°C in 90v/oN_2-10v/oH_2, Tempered at 200°C for 1 h.), 2%Nital/4%Picral, 200X original magnification complete transformation to martensite interspersed with areas of retained austenite. This more uniform structure results in the best mechanical properties and hardness. Care should be taken when selecting the chemistry for a sinter-hardening material, if the steel is made too hardenable excessive amounts of retained austenite will form, reducing strength and hardness.

These microstructural differences lead to significant changes in mechanical properties. For instance, the ultimate tensile strength of the 1.4Ni-1.25Mo-0.4Mn-2Cu-0.8C steel has a far superior strength level compared to the FLNC-4408, at the same or slightly lower densities as shown in Fig. 5.3(a). This is a direct result of the complete transformation of the microstructure to martensite. The same trend is observed with the FLC-4608. Apparent hardness increases both by complete transformation of the microstructure to martensite and by higher sintered densities. Figure 5.3(b) shows the superior compressibility and hardenability of the 1.4Ni-1.25Mo-0.4Mn-2Cu-0.8C results in higher apparent hardness values. The other materials tested would have to be cooled at a much higher cooling rate to raise hardness.

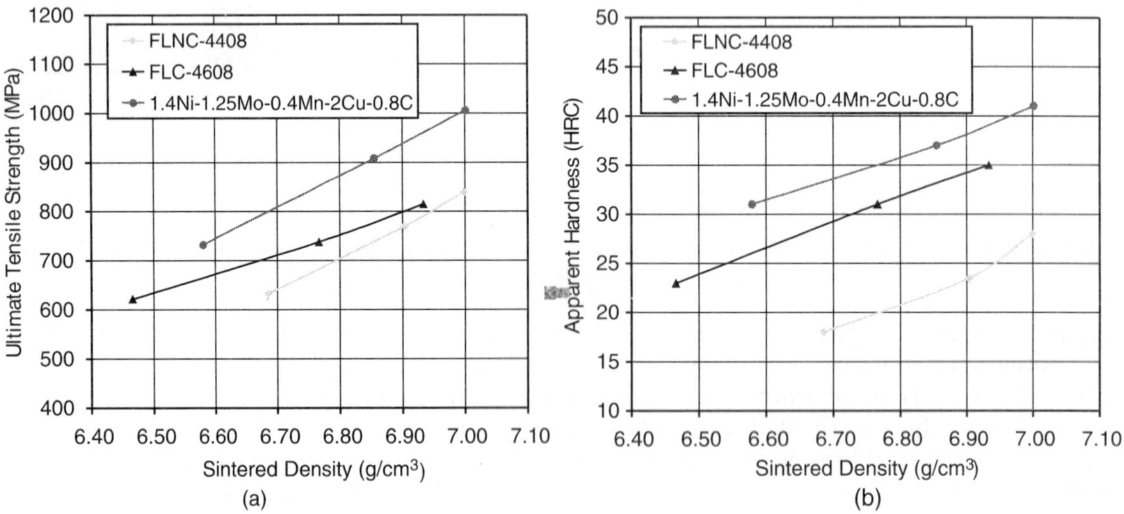

Fig. 5.3: (a) Ultimate tensile strength, and (b) Apparent hardness of select sinter-hardening materials. (Sintered at 1120°C in 90v/oN_2-10v/oH_2, Tempered at 200°C for 1h)

5.3 SINTERING FURNACE DESIGN

In addition to increasing alloy hardenability, the cooling rate from sintering temperature can be increased to produce higher levels of martensite in P/M steels. Specialized sintering furnaces have been developed with accelerated cooling sections to achieve higher cooling rates. Most commonly these systems use recirculated atmosphere gases that are cooled using a heat exchanger. Figure 5.4 compares two cooling profiles from the same furnace under the same loading conditions. The plot shows that with the accelerated cooling system activated, the cooling rate increased significantly. The highly linear line of the accelerated profile has an average cooling rate of ~65°C/min from 1100–200°C. The standard profile provides an average cooling rate of ~20°C/min with a long tail over the same temperature range.

Figure 5.5 shows a continuous sinter cooling transformation diagram for the 1.4Ni-1.25Mo-0.4Mn-2Cu-0.8C material [9]. Note that at the higher cooling rates shown on the diagram that the microstructure completely transforms to martensite [see text inset in Fig. 5.5(a)]. However, at lower cooling rates bainite and pearlite form, reducing hardness. Figure 5.5(b) shows the same diagram with actual sintering furnace cooling data overlayed. When accelerated cooling is used the knee of the bainite start curve is avoided, however, at the standard cooling rate the part will pass through the bainite/pearlite portion of the curve, resulting in a mixed microstructure.

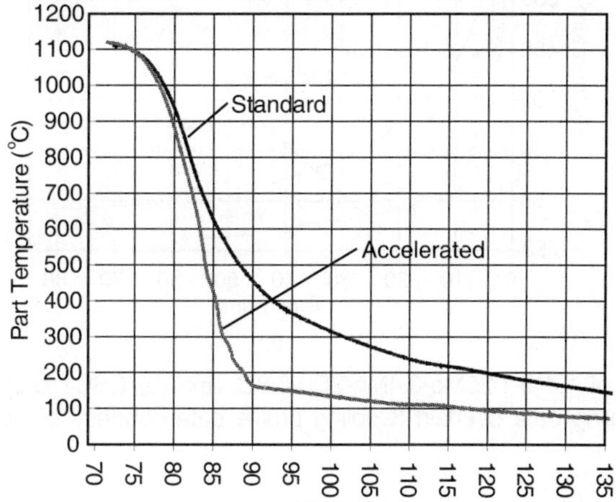

Fig. 5.4: Two cooling profiles from the same sintering furnace under the same loading conditions with and without accelerated cooling. (Cooling profile data courtesy of Abbott Furnace Company)

It should be noted that the cooling rate seen by the actual P/M part will change with furnace design, loading conditions, and part geometry. When developing a new sinter-hardened part it is recommended that a complete sintering furnace profile be performed with a thermocouple embedded within an actual part. This will produce a baseline to compare subsequent changes in furnace conditions and resulting cooling rates. This information can be used to optimize the final microstructure. The sintering furnace atmosphere should have a low oxygen content to minimize surface decarburization that leaves a soft surface layer on the sintered parts. This can be achieved by eliminating any leaks in furnace walls, properly balancing gas flow to stop air from being drawn into the furnace, and by using high quality gas sources and piping.

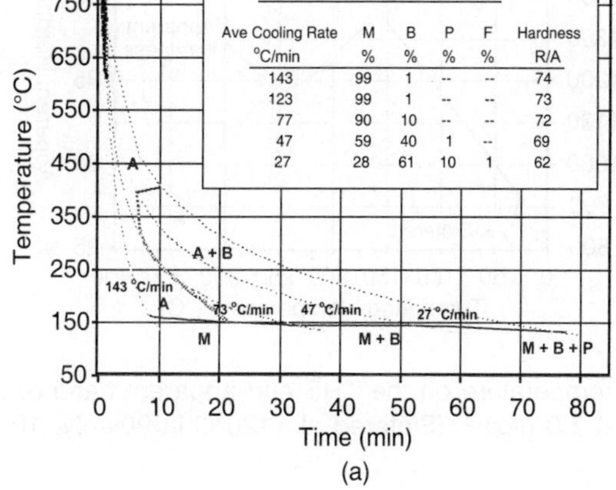

(a)

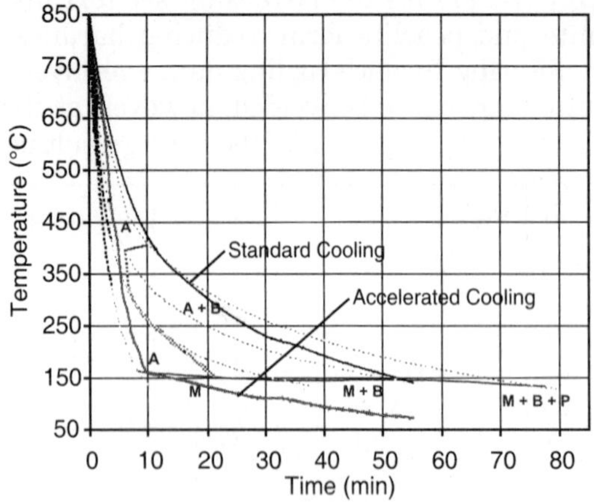

(b)

Fig. 5.5: (a) CSCT diagram of 1.4Ni-1.25Mo-0.4Mn-2Cu-0.8C, and (b) CSCT diagram of 1.4Ni-1.25Mo-0.4Mn-2Cu-0.8C with actual cooling data overlaid (Cooling profile data courtesy of Abbott Furnace Company)

5.4 TEMPERING

Tempering is also an important part of the sinter-hardening process. Tempering is employed as a stress relief that slightly softens the martensitic microstructure. This is done to reduce internal stresses that cause excessive notch sensitivity and brittleness; increasing the strength of the material. Figure 5.6 shows the transverse rupture strength (TRS) and apparent hardness plotted against tempering temperature for the 1.4Ni-1.25Mo-0.4Mn-2Cu-0.8C material. Note the much lower strength of the as-sintered material. The plot also shows that the tempering temperature can be adjusted to optimize strength or hardness. For the particular alloy shown a tempering temperature of ~200°C is generally recommended as a good compromise between strength and hardness.

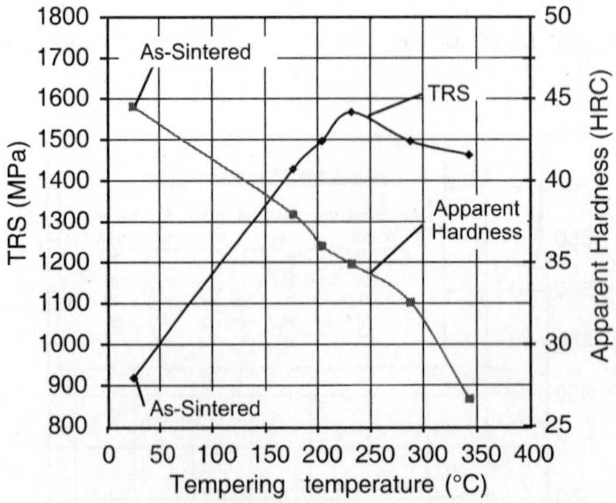

Fig. 5.6: Effect of tempering temperature on the TRS and apparent hardness of 1.4Ni-1.25Mo-0.4Mn-2Cu-0.8C at 7.0 g/cm³. (Sintered at 1120°C in 90v/oN_2-10v/oH_2)

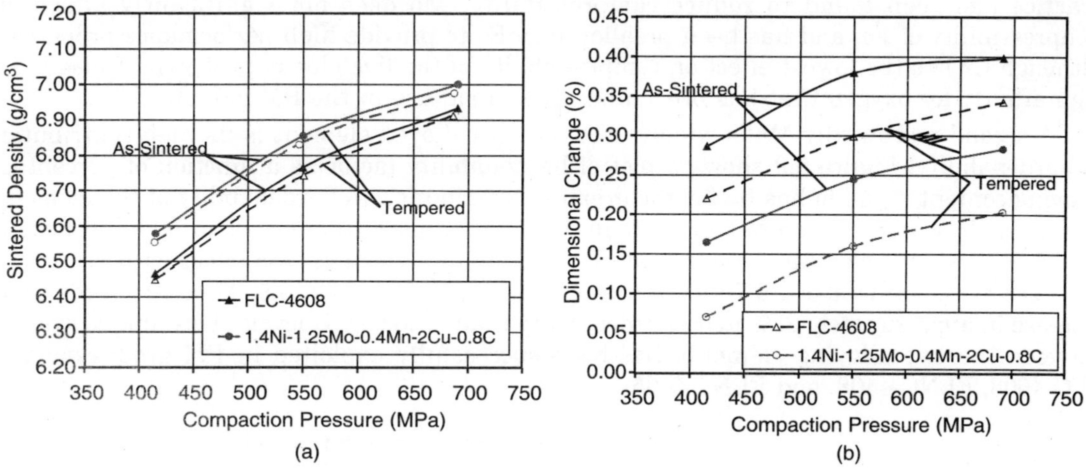

Fig. 5.7: (a) Sintered density, and (b) dimensional change of select sinter-hardening materials showing the effect of tempering. (Sintered at 1120°C in 90v/oN_2-10v/oH_2, Tempered at 200°C for 1 h.).

Tempering also has an impact on dimensional change. Figure 5.7 shows how tempering at a given temperature shifts these properties. For sinter-hardening materials admixed with Cu and graphite tempering generally reduces the growth from sintering. This is shown in Fig. 5.7(b) where for both alloys shown the dimensional change shifts downward by 0.15–0.18% over the range tested. This reduction in dimensional change is helpful in maintaining the density of the finished part and reduces dimensional variations. Change in tempering cycle will shift the dimensional change and the parts producer should keep the tempering conditions constant as possible to limit variation of the final part size. This shift will be unique for each alloy chemistry, oven, tempering profile, and loading condition and therefore must be optimized for each sinter-hardened part.

5.5 0.3% Mo PREALLOYED POWDER FOR HEAT TREATED APPLICATIONS

Powder metallurgy steels prealloyed with Mo have been widely used in the industry for the last two decades. The main benefits of Mo are its relatively minor effect on compressibility, a high contribution to hardenability, and an easily reducible oxide. Though conventional wrought alloys such as AISI 8620 and 4340 contain Mo levels less than 0.4 wt.%, a historically low and stable price structure has enabled ferrous powder manufacturers to successfully commercialize steels containing between 0.5 and 1.5 wt.% prealloyed Mo. Such alloys are typically premixed with elemental Ni and/or Cu, creating hybrid alloys that provide a desirable combination of compressibility and performance capabilities.

One of the main benefits that make Mo a favorable element for use in water atomized powder metallurgy (PM) materials is a relatively minor effect on compressibility. Figure 5.8 shows the effect on compressibility of various prealloyed elements at a compaction pressure of 415 MPa (30 tsi)(10). Prealloyed C is widely known to be deleterious to the compressibility of Fe powders, and instead is typically admixed in the form of graphite, which diffuses rapidly below conventional sintering temperatures. Mn, widely used in the wrought industry, also causes a significant decrease in the compressibility of Fe powder. A second downside to the use of Mn is its tendency to form a stable oxide at high temperatures [11,12]. Stable oxides are detrimental to the performance of ferrous PM materials. Ni has been successfully prealloyed in Fe, but in practice has been found to reduce compressibility. Mo does not significantly affect the compressibility of Fe, and has been prealloyed in Fe to provide high performance properties. Although Cr has the lowest effect on compressibility of the five elements shown, it also has a high affinity for oxygen that has prevented widespread use in the PM industry [12].

A second reason why Mo is widely used as a prealloyed element is its high contribution to hardenability. Figure 5.9 shows a plot of hardenability factor as a function of percentage alloying content [12]. Mo has by far the greatest contribution to hardenability of the elements shown, with nearly 1.5x more benefit than Mn and 2x more than Cr at an addition level of 1.0 wt.%. Perhaps even more interesting is that Mo has a synergistic effect with Ni. In the presence of Ni levels above 0.75 wt.% (designated by Mo*) in Fig. 5.9, Mo contributes to an increase in approximately 25% more hardenability than at lower Ni levels. This unique synergy between the two alloying elements has been successfully exploited in PM grades such as FLC-4608, FLNC-4408, and FLN2-4405.

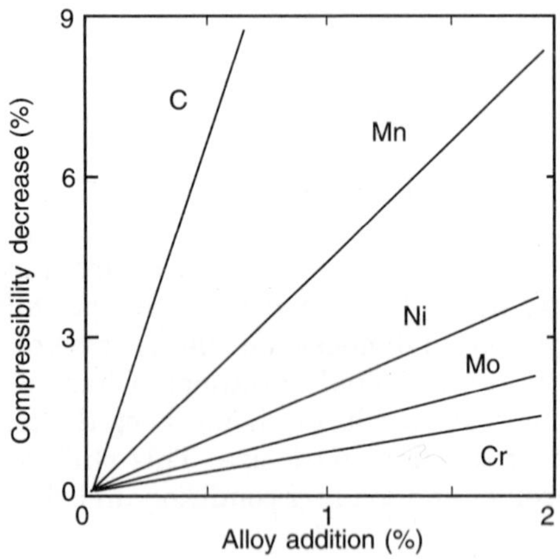

Fig. 5.8: Effect on compressibility of various prealloyed elements at 415 MPa (30 tsi)

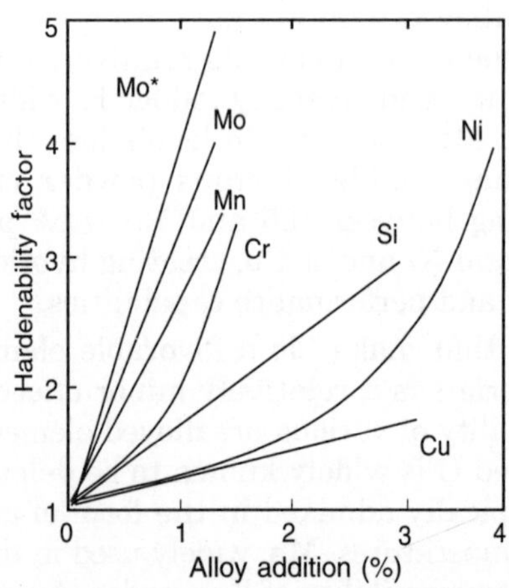

Fig. 5.9: Hardenability factors for various prealloyed elements [12]. Mo* is the hardenability factor of Mo when Ni is > 0.75 wt.%

A third benefit to using Mo as a prealloyed element is that it has an easily reducible oxide at conventional sintering temperatures such as 1120 °C (2050 °F) [14]. A diagram of metal/oxide equilibria in hydrogen for various elements is shown in Fig. 5.10. Cr and Mn both have a tendency to form stable oxides at elevated temperatures and as a result are not very commonly used in powder metallurgy steels. Mo, however, has an oxide that is fairly easy to decompose.

The three advantages described above, in conjunction with a low and stable price structure, have made Mo the most favorable choice for producing prealloyed powder metallurgy steels for nearly two decades. Over the past two years, however, Mo prices have soared above historic levels. As a result, powder metallurgy parts producers have been seeking to find alternative alloy systems and/or processing methods to provide high performance properties, particularly in the heat-treated condition.

A newly developed water atomized Fe powder, Ancorsteel® 30 HP, provides the benefits of Mo at a reduced level. Developed to help combat raw material prices, this steel contains nominally 0.3 wt.% Mo (Table 5.2) and is suitable for both heat-treat and as-sintered applications. With the surge of Mo prices, this composition offers parts manufacturers the ability to use a leaner composition for high performance applications while maintaining a desirable cost structure. Because Mo provides such a large improvement on hardenability compared to other elements (Fig. 5.9), lower Mo steel can still be used to provide high performance properties.

The objective of the current work is to illustrate the performance capabilities of an alloy with a reduced Mo content under production conditions. A comparative analysis of the recently developed 0.3 wt.% Mo steel with 0.85 wt.% Mo and Mo-free steels was conducted. Mechanical property data and metallography developed from the processing of a crankshaft sprocket are presented in the as-sintered, oil quenched and tempered, and induction hardened conditions at densities from 6.8 to 7.2 g/cm^3.

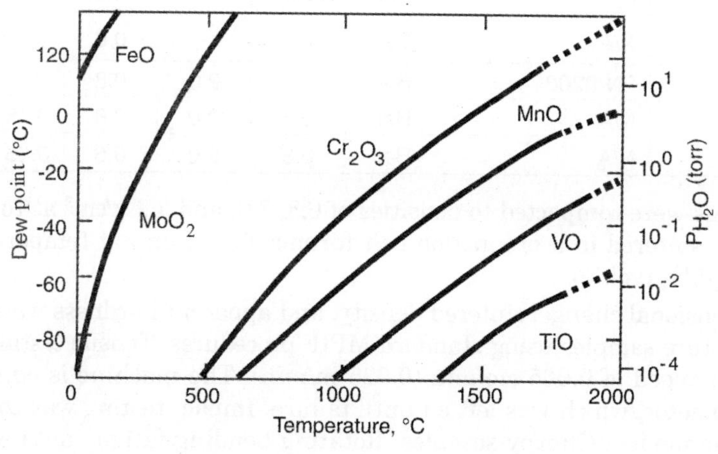

Fig. 5.10: Metal/oxide equilibria in hydrogen [14].

®Ancorsteel is a registered trademark of Hoeganaes Corporation.

Table 5.2: Nominal compositions of two prealloyed Mo steels evaluated in this study.

ID	Fe wt.%	Mo wt.%	Mn wt.%	C wt.%	O wt.%
Ancorsteel 30 HP	Bal.	0.35	0.1	< 0.01	0.1
Ancorsteel 85 HP	Bal.	0.85	0.1	< 0.01	0.1

5.6 EXPERIMENTAL PROCEDURE

Pilot scale premixes of FLN2-4405 and FN-0208 were made using water atomized Fe powders containing nominally 0.85 and 0.3 wt.% prealloyed Mo, respectively (Table 5.3). These powders are commercially available as Ancorsteel 85 HP and Ancorsteel 1000B. Two alloys were also made using a prealloyed base powder of 0.35 wt.% Mo which is commercially available as Ancorsteel 30 HP. Two additional premix versions of mix 2 were made with 0.35 wt.% additions of MnS and a second proprietary machining additive indicated as MA (mixes 5 and 6, respectively). Alloys contained 0.75 wt.% Lonza Acrawax C lubricant and Asbury 3203H graphite. Inco 123 was used as the Ni source.

Table 5.3: Nominal compositions of the alloys. All mixes contain 0.75 wt.% Acrawax C as a lubricant

Ix # Mix	MPIF Code	Fe wt.%	Mo wt.%	Ni wt.%	Gr wt.%	Additional wt.%
1	FLN2-4405	Bal.	0.8	2.0	0.6	-
2	N/A	Bal.	0.3	2.0	0.6	-
3	N/A	Bal.	0.3	-	0.6	-
4	FN-0208	Bal.	-	2.0	0.8	-
5	N/A	Bal.	0.3	2.0	0.6	0.35 MnS
6	N/A	Bal.	0.3	2.0	0.6	0.35 MA

Test specimens were compacted to densities of 6.8, 7.0, and 7.2 g/cm^3 at room temperature. The samples were sintered in a production belt furnace for 25 min at temperature at 1120 °C (2050 °F) in 95N_2-5H_2 (vol.%).

Per cent dimensional change, sintered density, and apparent hardness were measured from the transverse rupture samples using standard MPIF procedures. Tensile testing was performed using a crosshead speed of 0.065 cm/min (0.025 in/min). The machine is equipped with a 25 mm (1 in) extensometer, which was left on until failure. Impact testing was conducted at room temperature on unnotched Charpy samples. Rotating bending fatigue data were collected on as-sintered and Q&T samples of mixes 1 and 2 using a speed of 8,000 rpm at a nominal density of 7.0 g/cm^3. Specimens were machined from blanks processed under the conditions described above. The median fatigue endurance limit (50% FEL) was calculated by using the "staircase" method until there were both failures and runouts at a minimum of two stress levels.

5.7 RESULTS AND DISCUSSION

Green Density

Figure 5.11 shows a compressibility plot generated from cold compacted TRS samples for mixes 1–4. All four alloys demonstrate good compressibility, with green densities above 7.0 g/cm^3 at 550 MPa (40 tsi). Mix 2 had the highest compressibility, approximately 0.03 g/cm^3 higher than that of mix 1. The difference in alloy level between the two materials, a nominal 0.5 wt.% more prealloyed Mo in mix 1, correlates well with the expected loss in compressibility. As shown in Fig. 5.8, at a compaction pressure of 415 MPa (30 tsi), prealloying an additional 0.5 wt.% Mo is expected to reduce compressibility by 0.5%. With a green density of 6.84 g/cm^3 at this pressure for mix 2, a 0.5% loss in compressibility would equate to a loss of 0.034 g/cm^3, very close to what was observed experimentally for mix 1.

The 2.0 wt.% admixed Ni in mix 2 helps provide slightly higher compressibility over mix 3 because of the higher density of Ni relative to Fe (8.90 vs. 7.87 g/cm^3). Likewise, the compressibility plots of mix 1 and mix 4 are similar because the loss in green density due to prealloying 0.8 wt.% Mo in the former offsets the loss due to admixing an additional 0.2 wt.% graphite in the later. Graphite, with a density of about 2.2 g/cm^3, would be expected to cause a reduction in green density of 0.04 g/cm^3 in this case, a similar reduction to what Fig. 5.8 shows based on a prealloy Mo level of 0.8 wt.%.

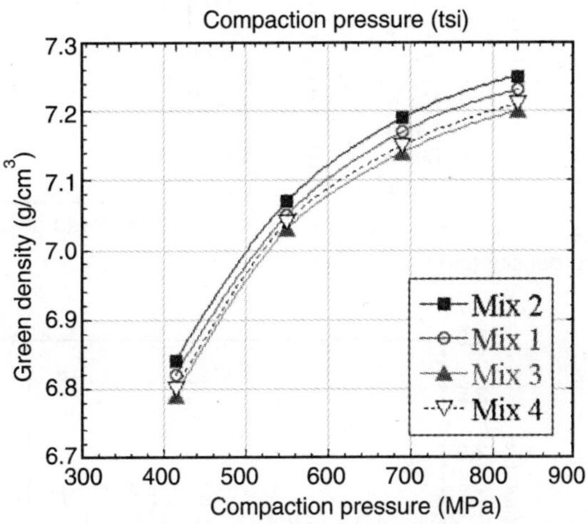

Fig. 5.11: Compressibility plot for the four alloys studied.

As-Sintered Properties

Plots of tensile properties as a function of sintered density are shown in Fig. 5.12 for the four alloys tested in the as-sintered condition. A summary of the remaining mechanical properties that were tested from 7.0 g/cm^3 green bars is shown in Table 5.4. As expected, the higher prealloyed Mo level in mix 1 provided a significant increase in strength and hardness. Likewise, this alloy showed lower elongation and impact performance than the leaner mix 2. Of interest to note is the dramatic effect of prealloyed Mo on performance compared to the presence of 2.0 wt.% admixed Ni. Mix 2 provides an increase in yield strength of 5–10% compared to a similar alloy with no admixed Ni, *i.e.*, mix 3. An increase in the prealloyed Mo level from 0.3 to 0.8 wt.% results in a yield strength increase of 25–30%. Because yield strength is directly related to microstructure, this property is attributable to the hardenability of the alloy. Referring to Fig. 5.12, an increase in Mo from 0.3 to 0.8 wt.% is expected to increase the hardenability factor much more versus an increase in Ni from 0.0 to 2.0 wt.%. This expected increase correlates well with the data observed in Fig. 5.12 and Table 5.4. A reduction in prealloyed Mo level from 0.8 to 0.3 wt.% increased ductility approximately 50%. As in the yield strength data, however, 2.0 wt.% admixed Ni had only a minor influence on elongation.The higher C level in mix 4 provided high strength performance despite having no Mo present. Although this material had relatively high as-sintered yield and tensile strength, it also provided a fair amount of ductility and impact toughness.Although the majority of PM parts that are fabricated from alloys such as these are typically heat-treated, these data were acquired to serve as a baseline to identify the relative hardenability levels of each of the compositions. Performance capabilities of the alloys under heat-treating conditions are shown in the following sections.

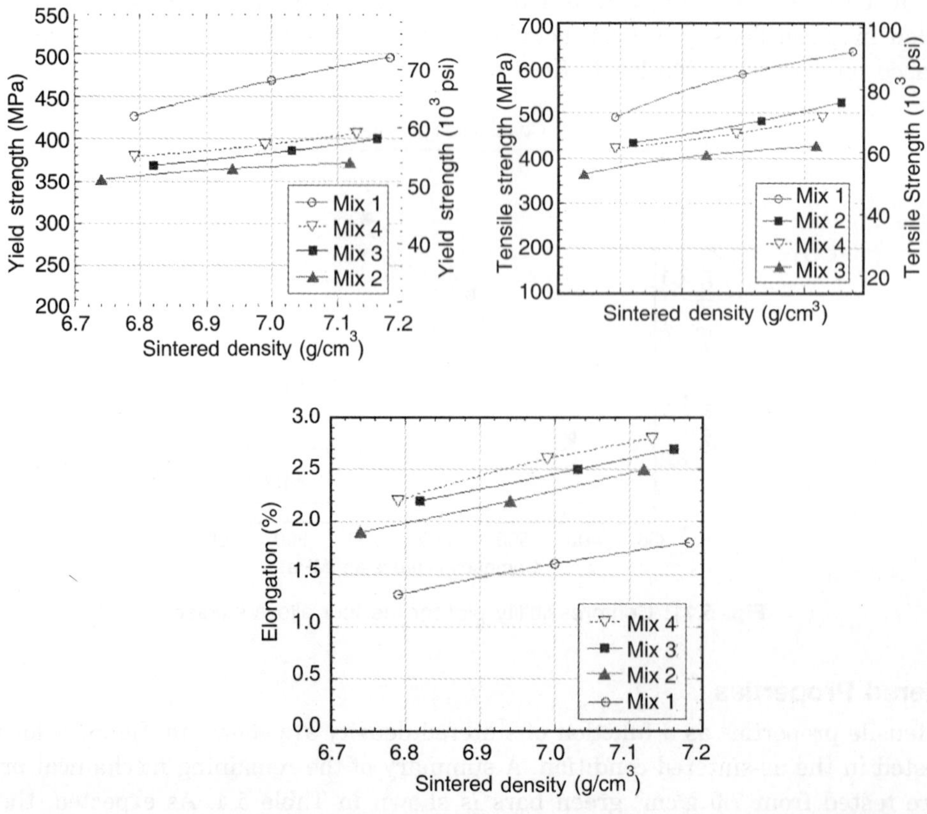

Fig. 5.12: Tensile properties for the four alloys studied in the as-sintered condition.

Table 5.4: Summary of as-sintered mechanical properties

ID	SD g/cm^3	DC %	TRS MPa (10^3 psi)	Hard HRA (HRB)	Imp. J (ft.lbf)
Mix 1	7.00	–0.04	1193 (173)	52 (84)	12 (9)
Mix 2	7.03	–0.08	1014 (147)	46 (74)	15 (11)
Mix 3	6.94	+0.17	772 (112)	38 (57)	9 (7)
Mix 4	6.99	+0.07	917 (133)	46 (74)	12 (9)

Q&T Properties

The four alloys evaluated in the as-sintered condition, as shown above, were also investigated under oil quench and tempering. Table 5.5 summarizes the Q&T properties for each of the materials. Fatigue data for mixes 1 and 2 are shown in Table 5.6. Fig. 5.13 shows plots of yield strength, tensile strength, and apparent hardness in as sintered and Q&T conditions.

Under Q&T processing, mix 2 provides very similar static and dynamic properties compared to mix 1 despite the lower Mo content. Mix 4 however, which contains no prealloyed Mo, had yield strengths that were approximately 15–30% lower than that of the other three mixes. It is evident that for Q&T processing, even a small addition of Mo provides significantly improved strength.

Important to note, however, is that these data were developed on test specimens with relatively small cross-sectional areas. For components with small distances between the case and core, alloys containing 0.3 wt.% Mo will perform similarly under Q&T processing to alloys containing 0.8 wt.% Mo. Larger components, however, will require a significantly higher hardenability in order to form martensite throughout the cross-section. In these instances, higher Mo contents may be required.

Table 5.5: Summary of Q&T properties for the four materials

ID	SD g/cm^3	DC %	TRS MPa (10^3 psi)	Hard HRA	YS MPa (10^3 psi)	UTS MPa (10^3 psi)	Elong %
Mix 1	7.05	–0.04	1476 (214)	72	848 (123)	1083 (157)	1.0
Mix 2	7.07	–0.07	1158 (226)	71	827 (120)	1096 (159)	1.1
Mix 3	6.98	+0.21	1269 (184)	70	800 (116)	972 (141)	0.9
Mix 4	7.03	±0.00	1358 (197)	71	710 (103)	779 (113)	0.8

Table 5.6: Fatigue endurance limit (FEL) values calculated from rotating bending fatigue testing for mixes 1 and 2 under different processing conditions

ID	Processing	90% FEL MPa (10^3 psi)	50% FEL MPa (10^3 psi)
Mix 1	As-sintered	218 (31.6)	225 (32.7)
Mix 2		191 (27.7)	197 (28.6)
Mix 1	Q&T	328 (47.5)	342 (49.6)
Mix 2		329 (47.7)	348 (50.4)

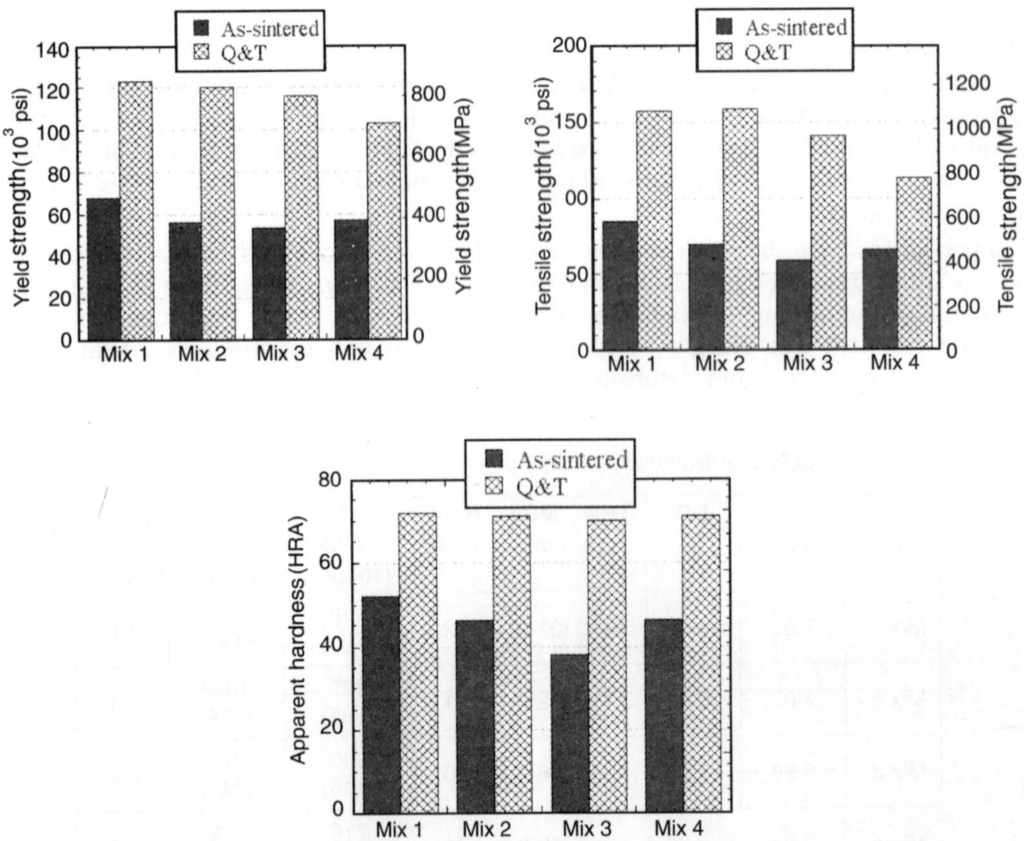

Fig. 5.13. Comparison of tensile properties and apparent hardness for the four materials

5.8 CONCLUSIONS

A comparative analysis of a recently developed 0.3 wt.% prealloyed Mo steel and more conventional 0.85 wt.% Mo and Mo-free steels was performed. A crankshaft sprocket was used as a prototype component. The study demonstrated that components with small cross-sections need only 0.3 wt.% Mo to provide desired strength and wear resistance properties in the heat-treated condition.

In heat-treat applications in which wear (*i.e.*, apparent hardness) and part strength are both critical, the more economically favorable Fe-0.3Mo-2Ni-0.6gr (mix 2) was found to have performance levels comparable to Fe-0.8Mo-2Ni-0.6gr (mix 1). In situations in which wear resistance is the main criteria, Fe-0.3Mo-0.6gr (mix 3) and Fe-2Ni-0.8gr (mix 4) are more cost effective alternatives that provide similar apparent hardness compared to alloys containing prealloyed Mo and admixed Ni. The comparative economic benefit of these two alloys is dependent upon fluctuations in Ni and Mo raw material prices.

Sinterhardening provides a lower cost option to quench and tempering. Sinterhardenable alloys are identified and processing details explained. The results suggest excellent strength and hardness can be achieved using Ancorsteel 737SH.

- Sinter hardening alloys should be chosen with the final properties in mind. The alloy should optimize hardenability for mechanical properties and compressibility for higher density while keeping the total alloy content as lean as possible to minimize cost. The microstructure should also be tailored to suite the needs of the required part performance, ensuring the proper degree of martensite transformation and if softer Ni rich areas are acceptable for the application.
- Sintering furnace design should employ a high cooling rate to improve the transformation of the microstructure to martensite. This can be achieved using a convection cooling system. The increased cooling rate can be used to employ a more compressible and leaner alloy or it can be used to increase furnace throughout with a more hardenable alloy. The sintering furnace should also employ an atmosphere with a low oxygen content to reduce surface decarburization.
- A tempering cycle should be employed when sinter-hardening. This increases mechanical properties and lowers dimensional growth. The changes in the tempering cycle will impact the final part size and keeping the tempering cycle constant will reduce dimensional variation and maintain proper part size.

REFERENCES

1. K.S.Narasimhan, "Recent advances in iron powder production for ferrous powder metallurgy applications.", Powder Metallurgy in Automotive applications, edited by P.Ramakrishnan, 1998, pp. 45–64.
2. K.S.Narasimhan and E, Boreczky New Transmission Technologies on Gear requirements for global powder metal industry, Society for Automotive Engineers, SAE 2006-01-0396.
3. Kalathur S.Narasimhan, Technology Advances for the growth of Powder Metal in automotive applications, Journal of Japanese Society of powder metal 2007, Vol. 54, No. 7.
4. James, W.B., "What is Sinter-Hardening", Advances in Powder Metallurgy and Particulate Materials – 1998, MPIF, Princeton, NJ, USA, 1998.

5. W.B. James, "High Performance Ferrous PM materials for Automotive Applications", Metal Powder Report 46(9), 1991.
6. F.J. Semel, N, Chawla, K.S. Narasimhan, "Binder Treated Analogs of Diffusion alloyed Compositions Based on Ancorsteel 150HP", Powder Metallurgy conference, MPIF, 2000, New York.
7. MPIF Standard 35, 2007, Published by Metal Powder Industry Federation, Princeton, NJ, USA.
8. M.C. Baran, A.H. Graham, A.B. Davala, R.J. Causton and C. Schade, "A Superior Sinter hardenable Material", PM conference1999, Vnacouver, B.C, Canada, Published by MPIF, Princeton, NJ, USA.
9. F.J. Semel, "Cooling rate effects on the metallurgical response of recently developed sinter hardening grade", PM conference 2002, Published by MPIF, Princeton, NJ.
10. German, R.M., "Powder Metallurgy & Particulate Materials Processing, MPIF, p. 195, 2005
11. N. Mootoka, N. Kunoishi "Strength and ductility of Mn-Cr sintered steel." Metal Powder Report, 1983, pp.629–631.
12. L.S. Siegl, P. Delabre, "Impact of oxygen on the microstructure and fracture morphology of Fe (Cr, Mo) PM steels". Powder metallurgy conference 2003, MPIF, Princeton, NJ, USA.
13. A.F. Deretana, J.V.Doane "Predicting hardenability of carburizing steels", Metals Progress 1971, p. 65.
14. R.M.German, "Powder Metallurgy Science" 2nd edition MPIF, 1994, p. 283.

❍❍

6

Development of Indigenous Sintered Metallic Valve Guide for High Speed Automobiles

G Babu[1], G Dheeraj[2] and Jiten Das[3]
1. Akhilesh Engineering, 2. Hyderabad, Indian Institute of Technology, Kharagpur
3. Defence Metallurgical Research Laboratory, Hyderabad, India

Abstract

Valve guide is a component which holds the intake valve and exhaust valve of any four stroke I.C. Engine. The exhaust valve, in particular, experiences high temperature exhaust toxic and corrosive gases. The intake and exhaust valves operate with a reciprocating motion at a very high frequency of nearly 10000 cycles per minute. The operating temperature is in the range of about 500–700°C. Generally for low speed engines, cast iron valve guides are used. However the oil retaining capacity of cast iron valve guides are low as the lubrication is forced. Moreover, oil consumption of cast iron valve guides is considerably high through drain and evaporation. Also abrasive and debris generated from cast iron contaminates the oil. Therefore, new materials with better mechanical properties are to be developed. Sintered valve guides made from blue dust iron have shown good promise as an alternative material for avoiding the above problems. Wear resistance is also one reason for the development of sintered valve guides. Sintered valve guides exhibit better wear resistance than cast iron under similar test condition. Sintered valve guides wore out less than 50 micron, whereas cast iron wore out up to 200 microns in our trials on an actual vehicle. Uniformity of microstructures is also an important factor for even and good machinability which is readily obtained by adopting the PM Technique.

6.1 INTRODUCTION

Cast iron valve guides have been commonly used in automobiles for long time. Cast iron has some features as guide material. It contains dispersed graphite flake formation for solid lubrication. Cast iron has good machinability. Machinability is very much required property for valve guide as the valve guide is finished to the correct size by reaming after press fitting into the cylinder head. The sintered valve guide contains oil in its pores and has excellent wear resistance. It will have good machinability also as it has to be reamed after pressing in the engine head block. Good machinability and good wear resistance are mutually contradicting properties. Sintered valve guide material contains hard phase matrix which imparts ware resistance to the material. The precipitated hard phases affect the cutting force of the final reaming operation. The present development of valve guide is to maintain a uniform micro structure, good solid lubrication, good machinability and better ware resistance.

6.2 EXPERIMENTAL

Prepartion of Components: Material Composition and Blending

The raw material powders are reduced sponge iron powders (micrograph in Fig. 6.1 shows the porous structure) from super concentrate blue dust. Average iron particle size less than 150 micron. The powder selected because of highly porous structure. The surface of the powder holds graphite as well as oil in its pores. Electrolytic copper powder, tin powder (average particle size of 45 micron), high purity graphite powder with ash content of less than 0.5% is used. Other fine powders used are Ferro phospurus, magnesium silicate, and manganese sulphide.

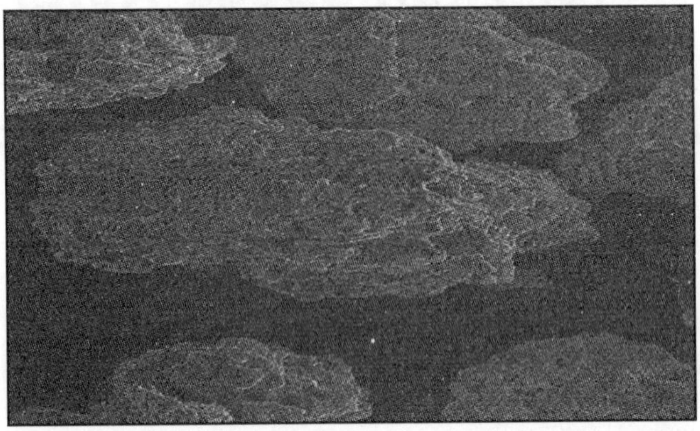

Fig. 6.1: Reduced sponge iron powder microstructure 150 micron

The powders were blended in the composition shown in Table 6.1, in a double cone blender. 6.1 Binder is added to prevent segregation of graphite and to keep the iron powder adhered uniformly with graphite. Zinc stearate used as lubricant. The powders are blended for 30 minutes.

Table 6.1: Powder blend composition

Chemical Composition (mass %)								
Fe	Cu	Sn	C	Fe_3P	$MgSiO_3$	MnS	Zn st.	Binder
Bal	4.50	0.50	2.40	3.00	0.60	0.15	0.45	0.35

Compaction

The powder mix compacted in deep cavity die on double acting automatic hydraulic press. The compaction tooling is specially designed and manufactured as the fill height is long. Bottom punch suction mechanism adapted for complete filling of die cavity. The powder compacted at pressure of 620 mpa. Dwell time of 3 seconds was set to stabilize the load among the green compact.

The final green compact size obtained was (dia) 4.45 mm × (dia) 10.35 mm × 38.00 mm (length).

Sintering

The compacts were sintered in hydrogen atmosphere in a pusher furnace at 1,060 degree C. for 45 minutes.

6.3 PROPERTIES EVALUATION OF COMPONENTS

Density

The density was measured by the Archimedean method. The part was cut into three sections as top, middle and bottom. The density of all three sections measured. The top and bottom portions density was 6.8 gm/cc., center portion density was 6.6 gm/cc.

Hardness

Hardness measured on Vickers hardness test machine with 5 kg load. The hardness varied between 68.8 and 96.5. The hardness variation was due to partial porosity.

Microstructure

The specimens were cut by a fine cutter and embedded in thermo set resin. After grinding and polishing microstructure observed under optical microscope and shown in Fig. 6.2

Fig. 6.2: Optical microstructure of sintered iron valve guide

Wear and Performance Evaluation

The practical approach of performance evolution selected to determine the ware out of conventional cast iron and sintered iron valve guides. The components was machined as per original part and replaced in two high speed four stroke 100 cc motor cycle engines. Both the valves are of same make used. Both the engines were operated under similar conditions for about 1000 hours. The engines were opened and valves and valve guides removed for dimensional studies. The dimensions before and after removal from the engine are tabulated. Results are shown in Table 6.2.

Table 6.2: Performance comparison

Parameter	Cast Iron Valve Guide	Sintered Iron Valve Guide
No. of Km of odometer	50,078	50,253
No. of hours operated (appx)	1000	1000
Average RPM	3500	3500
Oil used	2T	2T
Wear-out of valve stem	0.04 mm	0.02 mm
Wear-out of valve guide	0.06 mm	0.03 mm
Oil consumption	High	Low
Oil change required (average)	4,500–5,000 Km	6,000–6,500 km

Both engines on experimental basis run for 10 hours stopping oil to valves. Wear-out parameters tabulated in Table 6.3. The final product is pictured in Fig. 6.3.

Table 6.3: Wear-out parameters

Parameter	Cost iron valve guide	Sintered iron valve guide
No. of hours operated (appx)	20	20
Average RPM	3500	3500
Wear-out of valve stem	0.06 mm	0.03 mm
Wear-out of valve guide	0.185 mm	0.055 mm

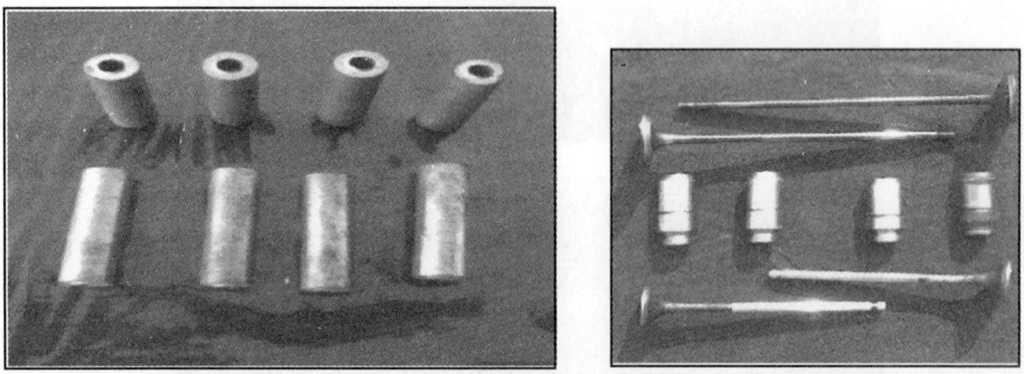

Fig. 6.3: The final products Sintered Valve Guide Blanks, Valve Guides

6.4 CONCLUSIONS

Use of Sintered iron valve guides renders lesser wear out of both the valve stem as well as valve guide as compared to the cast iron valve guide. In service Oil consumption in case of Sintered iron valve guides is lesser than that in case of the cast iron valve guide.

REFERENCES

1. Hayashi, K., Shikata,H., Ikenonue,Y., Ishii, K., ChiKahata, K., and Goto, G., Enhanced machinability of valve guides made from P/M Materials, MPIF ADVANCES in Powder Metallurgy & Particulate Materials 4, 1996
2. Funabashi, N., Endoh, H., and Goto, G., US–Japan PM Valve Guide History and Technology, Proceedings of the International Symposium on Valve train System Design and Materials, 1997.
3. Ikenoue,Y., Sato, R., Hayashi, K., and Chikahata, K., U.S. Patent 5,259,860.

❍❍

7

Development and Characterization of Iron Based Metal Matrix Composite for Brake Pads Applications

Ghazi A.A.S., Chandra K., Misra P.S., Chaturvedi A.K., and Asif M.
Metallurgical and Materials Engineering Department, IIT, Roorkee, India

Abstract

The present study deals with the investigations relating to development of iron based friction materials by newly developed 'Hot Powder Forging' technology and its dry sliding behaviour, reinforced with SiC_P and other additives. The dry wear test is carried out using a pin-on-disc tribo-tester at constant sliding speeds: 9 ms^{-1} under 6.1 kg/cm^2 load for sliding distance: 20 km as per ASTM G99 standard. The friction coefficient, temperature rise and noise level at contact point between test pin and disc is recorded and the wear loss of the test pins calculated from weight loss measurement. Sub-scale dynamometer test was performed to check its suitability for use in practical application. Formation of a compact transfer layer has been identified in the matrix region of the worn surface of the metal matrix composites.

7.1 INTRODUCTION

Most common types of friction brakes operate on the principle that friction can be used to convert the mechanical energy of a moving object into heat energy, which is absorbed by the brake. The essential components of a friction brake are a rotating part, such as a wheel, axle, disk, or brake drum and a stationary part that is pressed against the rotating part to slow or stop it. The stationary part usually has a lining, called a brake lining that can generate a great amount of friction yet give long wear life. The principal types of friction brake are the block brake, the band brake, the internal-shoe brake and the disk brake. All brake lining friction materials for a brake system should be designed to maintain stable and reliable friction force/coefficient at a wide range of pedal pressure, vehicle speed, temperature, humidity and others [1]. The performance (*i.e.*, safety, durability and comfort) of various brake situations is mainly controlled by the composition and microstructure of the brake shoe lining material [2].

A great deal of effort has been given to the development of multiphase composites for a friction material since a single material has never been successful to meet the numerous performance related demands [3]. More than 10 ingredients have been used to produce commercial brake friction materials, expecting that each ingredient provides beneficial roles for brake performance under different braking conditions. The ingredients in the friction materials are normally classified as binders, fibrous reinforcements, non-fibrous reinforcements and property modifiers, and more than 200 raw materials are currently used to produce commercial friction materials [4]. However, the type and the relative amount of ingredients in the commercial friction material are seldom found in the literature due to proprietary reason. Among various ingredients for property modifiers, solid lubricants and abrasives are known as two-key ingredients to achieve outstanding brake performance [5–7]. A few reports investigating the effect of solid lubricants on tribological properties of solid films coated on the metal substrates are available [8,9].

Compacting and sintering technology [10] is one such technology through which these brake pads are currently being made. However, it suffers from certain major limitations such as inadequate joining of friction element with backing plate, poor density levels achieved in friction element owing to limited application of pressure during compacting, poor thermal conductivity due to high levels of porosity in the product, poor strength due to segregation of the impurities along prior particle boundaries (PPB's), anisotropy in the strength of the product owing to preferred direction due of pressing, and, wide variations in final characteristics due to large number of variables involved. Besides above, cost of raw material and consumables along with heavy capital equipments makes this technique costly and only large scale production is possible where these costs are distributed in the large volume of production. In contrast to these limitations, the present technique can offer brake pads of much simpler chemistry but with improved performance on account of simultaneous application of pressure and temperature and with better control of variables. The technique does not involve any custom built equipments and is also likely to become economical even with small volume production [11].

7.2 EXPERIMENT

Friction materials investigated in this study were Iron-based metal matrix composites containing different constituents as indicated in Table 7.1. They are produced by 'Hot Powder Forging' technique. The powders are taken and mixed in double cone blender/ball mill in a specified sequence. This sequence involves firstly, the metallic powders like Fe, Ferro alloys, Cu, Sn, Al, P etc., were taken according to their weight percentages and then mixed in double cone blender for about 6–10 hrs depending upon its volume, flow rate, apparent density etc., then secondly, the coating of abrasive powders with lubricants for 2 hrs in ball mill with ratio of ball to material 1:0.5 in wet medium of ethyl alcohol. The coated powder in then dried in oven and thereafter, mixed with metallic constituents in ball mill for about 2 hrs. Thirdly, this mixture is then subjected to further mixing with lubricant in double cone blender for about 2 hrs [12]. The entire procedure of mixing is to obtain uniform distribution of constituents which have widely varying physical characteristics.The raw materials chosen in the present investigation have following size ranges as given in Table 7.2. By varying the raw materials three sample namely AG-1, AG-2, and AG-3 were prepared.

Table 7.1: Chemistry of friction materials

S.No.	Composition	% age
1	Fe	65 – 80
2	Cu	5 – 10
3	P	0.4 – 0.8
4	Graphite	6 – 8
5	SiC	3 – 4
6	Sb_2S_3	1 – 3
7	Others	4 – 9

Table 7.2: Sizes of powders employed

S.No.	Powder	Size Range	Source
1	Iron powder	–120 μm	Hoganas Industries Ltd.
2	Copper powder	–120 μm	Electrolytic
3	Graphite	–200 to +150 μm	Natural Crystalline
4	Ferro-phosphorus	–45 μm	Commercial grade
5	Tin powder	–75 μm	Atomized
6	Silicon Carbide	–180 to +150 μm	Chemical grade
7	Calcium Sulphate/ Barium Sulphate/ Antimony Trisulphide	–45 μm	Chemical grade

The powder mixture is then filled inside a steel capsule as shown in Fig. 7.1. Hydrogen flow is restored thorough the capsule and it is heated in the temperature range of 1000°C–1100°C for about 1 hr. where the powder is effectively cleaned of any reducible impurities. The other end of the steel capsule serves as outlet for the hydrogen which burns during heating. The hot steel capsule containing powder mixture is then transferred to a channel die where it is hot forged using friction screw press. The speed of the pressing is of the order of 300 mm/sec. and the pressure employed during forging is of the order of 2.5 metric tons/ sq. cm. During forging the supply of hydrogen is put off as high density consolidation has been effected where by the produce becomes resistant to further oxidation. Furthermore, the capsule cover provides protection from oxidation in subsequent processing. The overall reduction in the cross sectional area of the powder during forging is about 50 to 60% having forged density of the consolidated slab to about 85% of theoretical. The forged slab is subjected to homogenization treatment of about 1100°C for 1 hour. and it is followed by hot repressing in a hydraulic press to improve the density of forged slab. The homogenized slab is subjected to annealing treatment after repressing by holding it in the furnace kept at 700°C for 5 hrs. Thereafter the slab is cooled in the furnace itself. The annealed slab is then machined to give final shape of the brake pad retaining flattened steel capsule as backing plate. The whole sequence of processing described above is summarized in Fig. 7.2 as process flow chart.

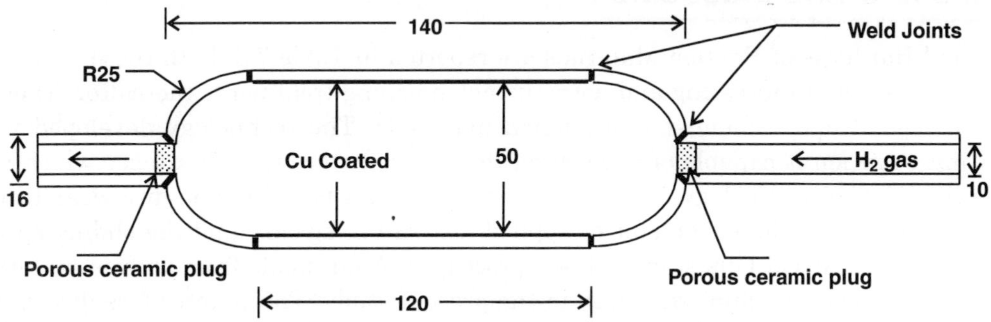

Fig. 7.1: Steel capsule

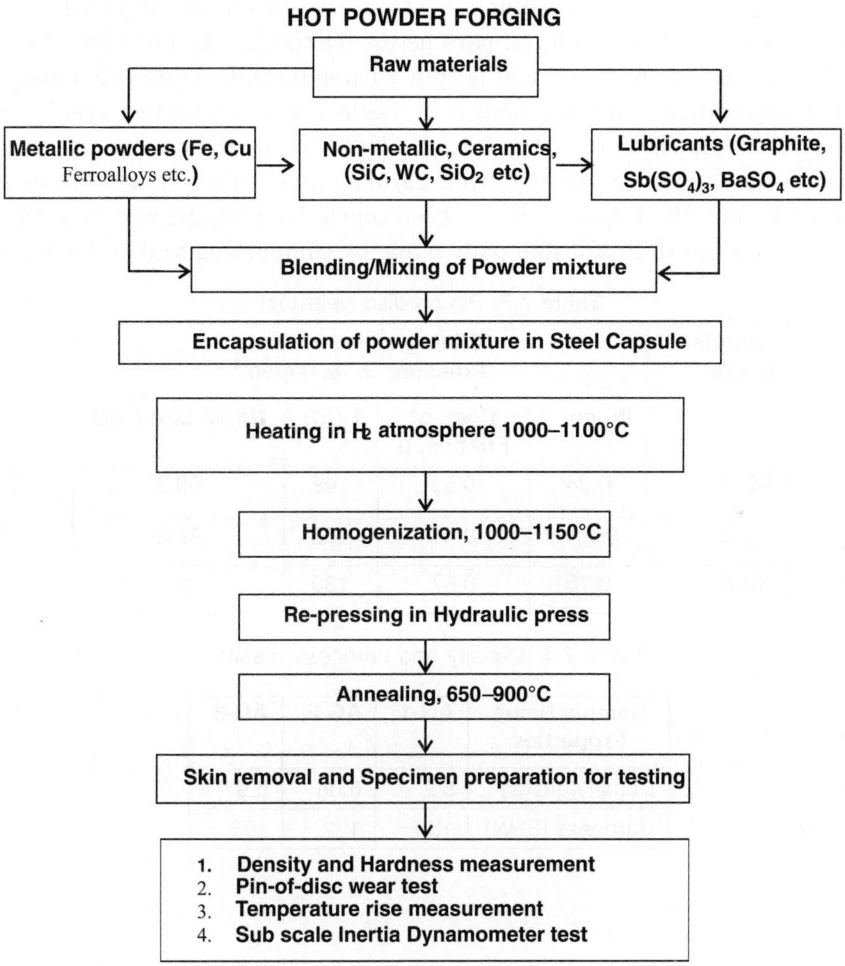

Fig. 7.2: Process flow chart

7.3 RESULTS AND DISCUSSION

Density and Hardness of Friction Materials are reported in Table 7.3. Both these parameters depend upon extent of repressing and normalizing/annealing treatments thereafter. They also indirectly depend upon, chemistry of friction materials. The technology developed in the present investigation is capable of developing these parameters in wide ranges which is not possible in sintering technology. An initial laboratory level based pin-on-disc wear test has been performed under the parameter of applied load of 6.1 kg/cm^2, and the sliding speed of 9m/s, according to ASTM G99 standard test procedure. A pin made from friction material of having cross-section 7×7 mm^2 was used in the present study. The counter face disc is made of cast iron. The noise generated during the wear process is also recorded by placing a microphone based noise level meter in decibels (dB) at a distance of 25 mm from the point of contact between friction pin and rotor disc. Rise in temperature during the wear test is also measured by placing thermocouple 5 mm away from the initial wearing surface. The results are tabulated in Table 7.4. The test input parameters for the AN-32 aircrafts rotor applications are shown in Table 7.5. The required test output parameter values are calculated accordingly for AN-32 rotor application and are shown in Table 7.6 [14]. The test results of Sub-scale dynamometer are summarized in Table 7.7 for AN-32 rotor application; the run-down time are nearly in the same range as that of output parameters (Table 7.7) and the wear is slightly lower in these pads than the required. It has been concluded that the performance of proposed brake pads is better than that of brake pads made by compacting and sintering technology.

Table 7.3: Pin on disc wear test

Sample Name	Sliding speed = 9m/s Pressure = 6.1kg/cm^2			
	W, gm	Coef. of Friction, μ	T (°C)	Noise Level, dB
AG-1	0.04	0.53	149	99.3
AG-2	0.174	0.41	163	92.6
AG-3	0.15	0.57	130	95

Table 7.4: Density and hardness results

Sample Name Properties	AG-1	AG-2	AG-3
Density(gm/cc)	5.2	6.06	5.9
Hardness (BHN)	132	122	136

Table 7.5: Test input parameters brake pad contact area: 25.8 cm^2

Applications	Test Parameters		
	Kinetic energy, kgfm	Brake Speed, rpm	Brake Force, kgf
AN 32 Rotor	17300	1000	150

Table 7.6: Required test output parameters

Application	Test parameter					
	RD Time, sec		COF		Maximum Wear (for 50 cycles)	
	Min	Max	Min	Max	in grams, gm	in thickness, mm
AN 32 Rotor	6	12	0.18	0.40	14	1.25

Table 7.7: Test results for AN-32 rotor application

Sample	50 Cys.	RD revl.	RD time, sec.	COF	Mean Torque, kgfm	Peak Torque, kgfm	Mean Drag, kgf	Peak Drag. kgf	Wear			
									Thick.,mm		Wt., gm	
									Pad1	Pad2	Pad1	Pad2
AG-1	Max	120	14.4	0.28	27	65	148	359	0.11	0.07	1	1
	Min	96	11.6	0.22	21	43	119	240				
	Avg	107	12.6	0.25	24	52	133	290				

7.4 CONCLUSIONS

1. Any shape and sizes of brake pads is possible to produce for variety of vehicles.
2. The technology completely eliminates the problems of joining between backing plate and friction element on account of simultaneous application of pressure and temperature.
3. Higher density levels can be achieved by Hot Forging Technique.

Acknowledgements

We sincerely acknowledge the help and facilities provided by the Foundry and Forge Division, HAL Bangalore in carrying out the testing on sub-scale dynamometer in their Powder Metallurgy Shop.

REFERENCES

1. S.J. Kim, H. Jang, Friction and wear of friction materials containing two different phenolic resin reinforced with aramid pulp, Tribol. Int. 33 (2000) pp.477–484.
2. W. Österle, M. Griepentrog, Th. Gross, I. Urban, Chemical and microstructural changes induced by friction and wear of brakes, Wear 251 (2001) pp.1469–1476.
3. S.J. Kim, M.H. Cho, D.-S. Lim, H. Jang, Synergistic effect of aramid pulp and potassium titanate whiskers in the automotive friction material, Wear 251 (2001), pp.1484–1491.

4. M.G. Jacko, S.K. Lee, Brake linings and clutch facings in Kirk–Othmer Encyclopedia of Chemical Technology, Vol. 4, Fourth ed., 1992, pp.523–536.
5. H. Jang, S.J. Kim, Wear 239 (2000), pp.229–236.
6. L. Gudmand-Hoyer, A. Bach, G.T. Nielsen, P. Morgan, Wear 232 (1999), 168–173.
7. L.C. Lipp, Lubr. Eng. 2 (1975), pp.574–584.
8. W.J. Bartz, J. Xu, Lubr. Eng. 43 (1987), pp.514–521.
9. M.N. Gardos, Tribol. Trans. 31 (1988), pp.214–227.
10. Dixon R.H.T., Clayton A., 'Powder Metallurgy for engineers', Machinery Publishing Co., Ltd., London, 1971, pp.166–183.
11. Dutta D., Mohan G., Chatterjee B., Krishnadas Nair C.G., 'Development of sintered metalloceramic friction material for the wheel brakes of a military transport aircraft',published in, 'Composites: Science and Technology (ISBN 81-224-1251-3)' edited by R.C. Prasad, P. Ramakrishan, New Age International (P) Ltd., New Delhi, 2002, pp. 94–114.
12. Lenin D. Singaravelu, PhD. Thesis 'Development of Friction Materials Through Powder Metallurgy', 2007.

❍❍

8

Expected Changes in PM Part Compaction

Norbert Nies, SMS Meer GmbH Germany

Abstract

The PM industry today is faced with both economical and ecological changes. From the point of view of SMS Meer, new trends are essentially to be seen in the fields of drive and energy technology, and generally in the areas of part size and geometry, material mixtures and composite compacting. The questions that arise necessitate lasting solutions covering practically all important areas of PM compaction using modern CNC presses. The presentation focuses on a few selected compacts with differing levels of complexity developed to meet specific demands. The process-oriented design possibilities illustrated highlight individual solutions that have been developed quickly and flexibly. The listeners are also given ideas for developing creative answers for their own applications.

8.1 INTRODUCTION

This presentation describes based on the advancements of CNC/CPA technology and even the combination of withdrawal-and counter pressure-process to fulfil the demands of said expected changes in PM part compaction. Through the use of the latest CNC and adapter technology it is possible to produce even highly complex sinter parts to meet the highest quality demand, crack free, with high density and very close tolerances in a high production rates. If we consider sintered compacts today, we observe that there has been a change in the form and size of the parts, with demands of more than 7.1 g/cm^3 and dimensional tolerances of 0.04 mm being no longer unusual. The examples (Fig. 8.1) below show sintered compacts as of today produced or developed for production on hydraulic CNC powder presses with highest demands at thin walls, long top section drop length, density, size precision, force deviation between punches. Some of them have been awarded prices for their exceptional quality.

Fig. 8.1: Examples of compacts as of today

Starting from these components it can be observed that the sintered parts have developed in parallel in different directions (Fig. 8.2).

Fig. 8.2: PM part development in different directions until today

On the one hand, the densities of the green parts could be significantly increased thanks to new powders, whereby we should point out that an increase in density from 7.0 g/cm³ to 7.3 g/cm³ is linked to an increase in the press force of 50%. On the other hand, the required press force increased due to the ever-growing part diameters that in turn led to ever-larger presses with higher press forces and larger numbers of press levels. Furthermore, the complexity of the parts increased visibly. The sinter parts not only became higher and had more levels requiring extreme powder transfer movements with the corresponding strokes within the tool, but the punch planes proper vary to such an extent in size and form that particularly during withdrawal and during force relief, controlled punch movements are necessary in order to avoid cracks.

8.2 INFLUENCES

If we take this achieved standard with respect to punch level–force–density–quality–as an equivalent target as the basis for the following considerations, then from the point of view of the aspects energy, emissions, economics and benefits, the following two paths for the future can be identified. Of course these aspects can and must first be analysed and evaluated individually with respect to the PM part production, something that is only possible to a limited extent here, and on the other hand they form large cut-sets and inter-dependencies so that these are summarised only in their key points. Energy, or more precisely the energy consumption during the production of sinter parts, is considered very critically in this context. It is hereby fundamentally correct to analyse all the individual process steps, but we also have to consider whether possibly the parts have to be produced more near net shape in order, *e.g.*, to reduce mechanical finishing. The extent to which material modifications or the use of dual composite material parts even in the press process can save other working steps will therefore gain in significance. Last but not least we have to consider the tooling times for the machines, the time scale for the practical development tests and the productivity of good parts–in other words, reducing scrap to practically zero. As far as the energy consumption during the compacting of powders is concerned, powder presses with part-oriented power adaption are already being built today.

Emissions and automotive are directly linked and will remain inseparable also in the future. The fact that emissions have to be reduced worldwide while at the same time paying critical attention to the energy consumption and energy resources is reflected not least by the current situation in the automotive industry. It remains to be seen which drive system will be used in future, or which concepts will establish themselves for the various drive types. There is a high probability, however, that there will be three classes or sizes of drive types. Compact car – electric drive; standard car–hybrid drive; truck–internal combustion engine Economics in combination with automotive is an equally interesting field. Here the focus is not on the price, but on the consideration of which concept is technically feasible in large quantities with a high energy yield and ultimately affordable. Powder metallurgy has a great chance of contributing directly to the solution.

Benefit is a business principle also for the future. We can certainly discuss the level of the benefit, but not at any price. Technically, new solutions have to be considered in particular also from an ecological point of view, while at the same time being profitable. When we now consider that ultimately the same key words are of importance also in other branches of business and industry, then it is perhaps easier or even imperative to expand our technology and know-how also to other fields of application. This potential has to be utilised and already offers fields of application as described here.

8.3 PERSPECTIVES FOR THE FUTURE

The know-how and the influences described allow two general trend directions for the production of PM compacts to be identified. Version A (Fig. 8.3) shows that the number of press axes and punch levels and the density can be further increased for various applications, while the forces will drop with new parts of decreasing size. The demand on quality, however, will continue to increase for all parts.

This means in detail: Size of compacts will not become bigger, but more complex with up to 4 top and 7 lower levels and therefore the punches proper will become smaller size of compacts will tend to become smaller, but multi-level compacts with the same amount as of today. New part shapes will be designed whereby more non-symmetrical forms are requested increased demand for higher density of compacts, based on new powder mixes weight of compacts will be reduced by the use of *e.g.*, Al powder much higher demands on quality which will affect the compaction process first.

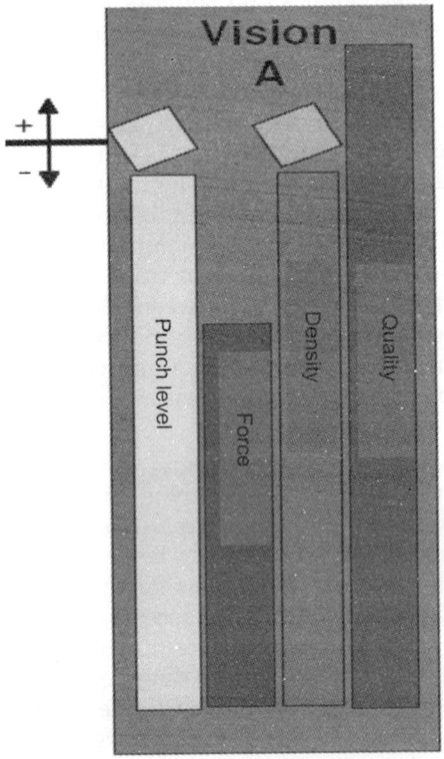

Fig. 8.3: Trend version A

Fig. 8.4: PM part examples (Courtesy: Schunk Sintermetall) (Courtesy FMS Corporation) (Courtesy: Mitsubishi materials PMG)

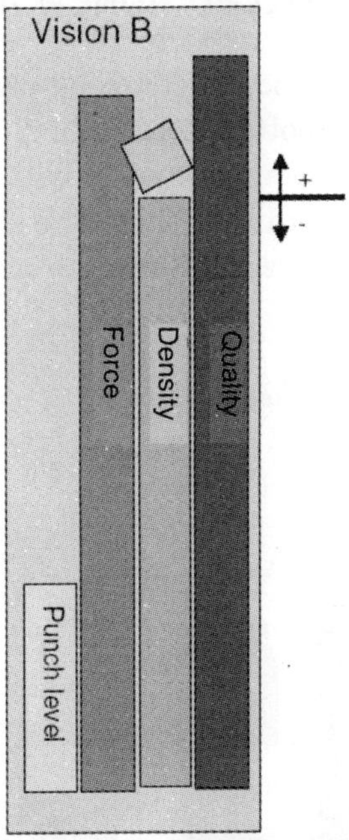

This means in detail:

New applications for PM parts such as

e.g., Fuel cell solutions
• Mobile phone/car
• Stationary applications

e.g., Electric drive components

Composite material solutions such as

e.g., Oil pump gear (Fig. **8**.6)

e.g., Synchronizer ring (Fig. **8**.7)

Applications based on new materials

Fig. 8.5: Trent version B

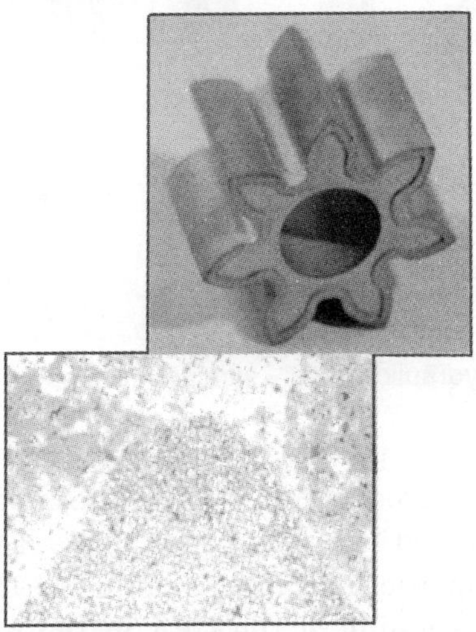

Fig. 8.6: Oil pump gear

Fig. 8.7: Synchronizer ring

Version B (Fig. 8.5) shows a quite different trend in which the number of press axes and punch levels is greatly reduced, but at the same time the press forces increase. The influence of density and quality is evaluated here in the same way as for Version A.

8.4 CNC HYDRAULIC PRESS WITH CPA-CONTROLLED PUNCH ADAPTER

The philosophy behind our technology is to allow sintered compacts to be produced using a powder press/CPA combination in which several top and bottom hydraulic punches perform a cycle and all punches can travel to and stop at their compacting position with 100% accuracy without any mechanical stop (Fig. 8.8). In order to achieve this, the latest machine engineering has been linked to the best valve technology, precision measurement technology and a sensitive high-speed CNC controller (Fig. 8.9–8.11).

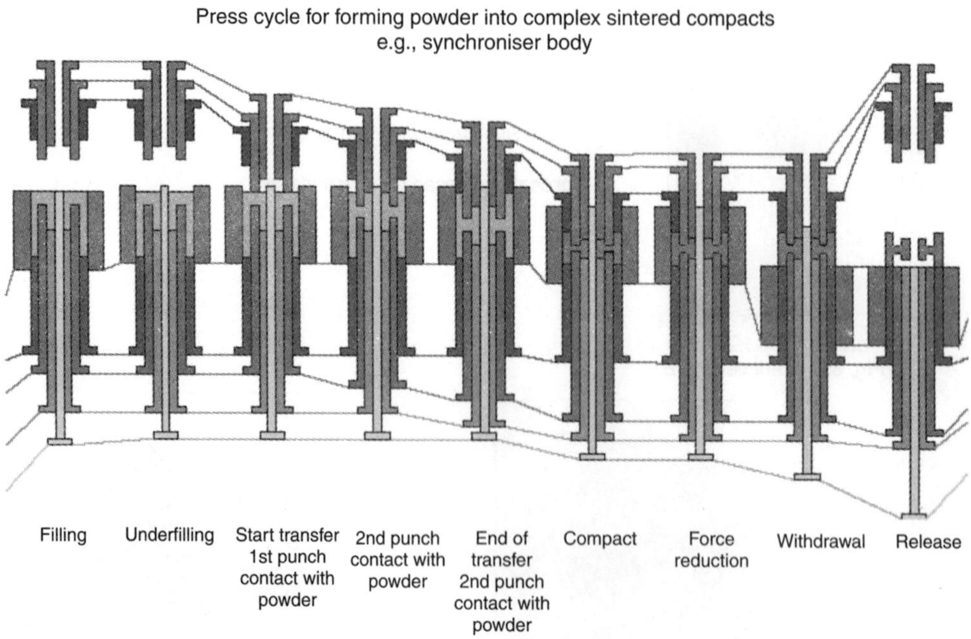

Fig. 8.8: Cycle diagram of CNC hydraulic press - synchronizer hub

An appropriately dimensioned hydraulic power pack is provided for the optimum use of the press/CPA combination to achieve maximum production of sintered compacts per unit of time, and supplementary facilities are integrated into the press that also promote this maximisation of production, such as *e.g.*, automatic hydraulic coupling of CPA and press, automatic lifting/lowering device for quick positioning of the CPA in the press, transport device on the CPA for rapid adapter transport, and tool changing and parking station connected to the press by rails for quick changing of the adapter. A control system with simple and clear dialogue for the input and storage of machine and tool data further enhance the optimum utilisation of the press. Controller specifications help in quality monitoring, inspection of new parts and troubleshooting.

Fig. 8.9: Press front side with control panel and safety equipment CPA installed in press

Fig. 8.10: Press rear side with filler system, powder platform tool changing and parking station

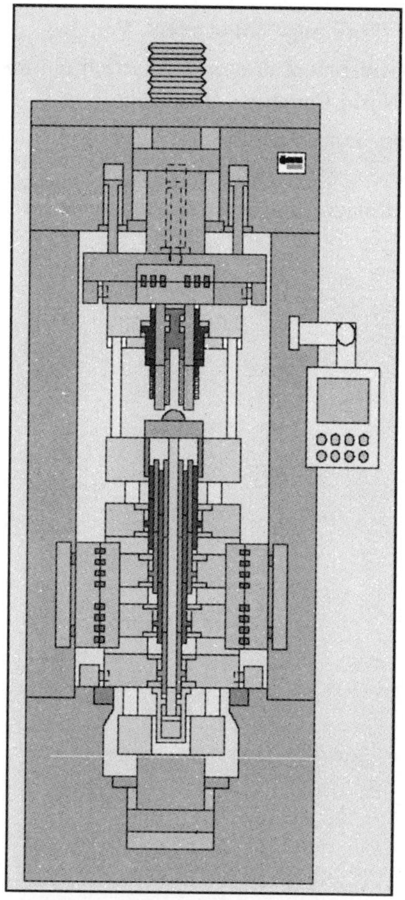

Fig. 8.11: Schematic representation of a hydraulic CNC powder press with CPA and supplementary equipment

8.5 CONCLUSIONS

In the final summary we can say that the CPA press technology is fully capable of meeting the new challenges with which we are faced today, also due to the economic crisis, and is waiting for effective utilisation in the form of innovative applications.Today the time is right to create new compacts which will bring all of us ahead in the PM field.

REFERENCES

1. EPMA PM Training Course 2008, Vol. 1.
2. Nies,N., Development of new compaction methods and process progressions with CNC–CPA Systems, 1996 PM World Congress, Washington.
3. Pulverpresse mit etagenförmiger Grundplatte im Adapter, Deutsches Patentamt DE 42 27 640 A1, 1992.
4. MPIF-PM Design Center-Award Winning Parts.

❍❍

9

High-Speed Presses Increased Economy in Powder Compacting

Hubert Plank, DORST Technologies, Germany

DORST Technologies has developed a series of high-speed presses making it possible for PM producers to challenge the extreme cost pressure in the present difficult economic situation. The presses ranging from 6 t to 70 t are already time-tested and have been successfully operating since a number of years. The quality of the compacts pressed on high-speed presses has been significantly improved in many aspects as compared to standard presses. The production output could be increased by the factor 2 to 2.5 in comparison with standard presses. 24/7 operation is possible Using automatic handling systems also provided by DORST Technologies.

9.1 INTRODUCTION

PM producers have to face the facts resulting from the present unpleasant global economic development. In powder compacting especially simple shaped parts are under extreme cost pressure. A few years ago, DORST already started to develop mechanical high-speed presses based on the well-known mechanical TPA presses. These high-speed presses enable increased output coupled with improved quality.

The presses are:

Fig. 9.1: TPA6/2S

Fig. 9.2: TPA 15/4S

Fig. 9.3: TPA 50/4S

Fig. 9.4: TPA 70S

In order to achieve a perfect performance of the high-speed presses, a number of improvements have been necessary. First of all the kinematics had to be improved.

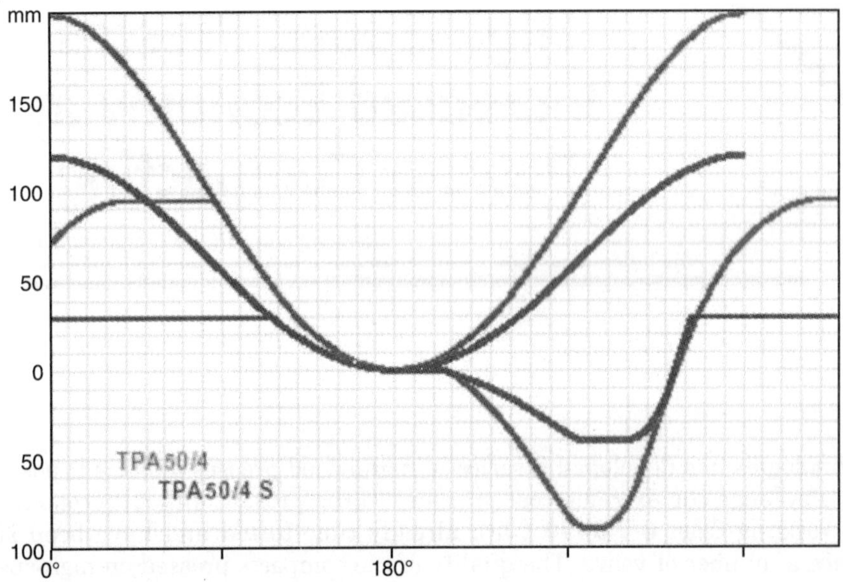

Fig. 9.5: Diagram of upper punch and die movement

Smooth movements, soft acceleration and deceleration are the keys of high-speed presses. Cams for upper punch movements die movements and filler movements had to be redesigned. Special attention had to be paid to the filling process. Results were beyond expectations. With increased stroke rates filling behaviour improved, weights of the compacts became more constant.

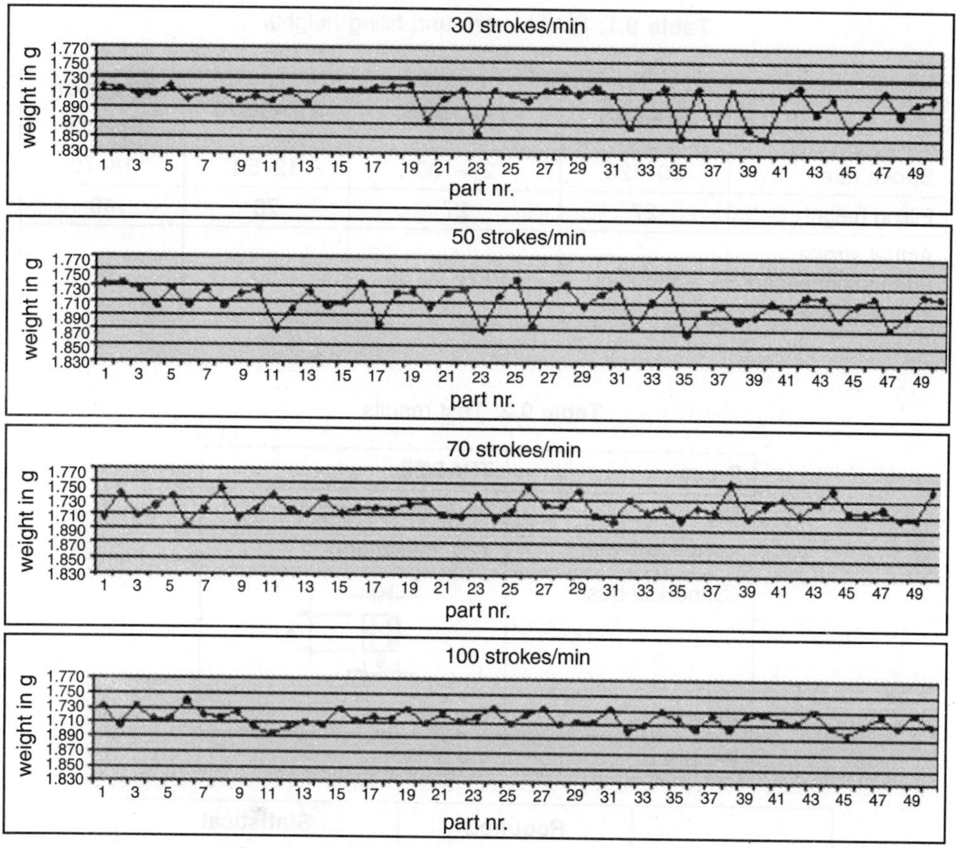

Fig. 9.6: Weight of compact at various stroke rates

This is in contradiction to standard experience in practise.

9.2 POSSIBLE EXPLANATION

Due to the high speed of the filling shoe, powder at no times gets to rest anymore.

Flowability increases and filling becomes more constant. This tendency we have noticed with different powders, ceramic powders and as well as iron powders.

Thanks to improved filling not only a more even density, more even weight and height but also a reduced radial run-out is achieved. In a word, the quality of the pressed compact improves.

Table 9.1: Stroke rates and filling heights

	TPA6/2S	**TPA15/4S**	**TPA50/70**	**TPA50/70 Special**
Stroke rate	30–120	25–100	12–50	18–70
Filling height	27	27	70	40
Actual stroke rates experienced in production	Approx. 80	APPROX. 80	50	60

Table 9.2: Test results

Press	**TPA6/2S**
Die	4 fold (4 openings)
Stroke rate/min.	120 (maximum)
Cylindrical bush	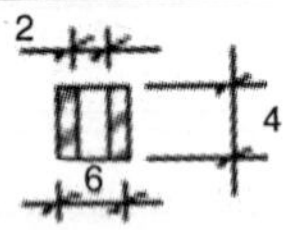
Powder	Copper
Weight	8.9 gms

	Required Tolerance	**Statistical C_m Value Reached**
Radial run out	± 0.02 mm	2.43
Height	± 0.07 mm	2.63
Weight	± 0.13 gr	2.83

Note: A C_m value of 1.67 is generally accepted in production

$$C_m = \frac{\text{upper limit} - \text{lower limit}}{6\,S} \qquad S = \text{Standard deviation}$$

Contd...

Table 9.2: Test results

Press	**TPA 15/4S**
Die	1 fold (1 openings)
Stroke rate/min.	60
Cylindrical bush	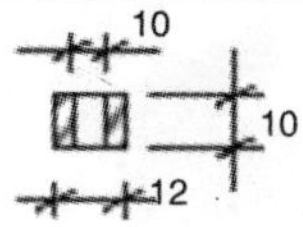
Powder	Sint B
Weight	3.00 gr

	Required Tolerance	**Statistical C_m Value Reached**
Radial run out	± 0.013 mm	1.89
Height	± 0.1 mm	4.36
Weight	± 0.1 gr	3.49

Note: A C_m value of 1.67 is generally accepted in production

$$C_m = \frac{\text{upper limit} - \text{lower limit}}{6\,S} \qquad S = \text{Standard deviation}$$

Press	**TPA 15/4S**
Die	1 fold (1 openings)
Stroke rate/min.	60
Flange bushing	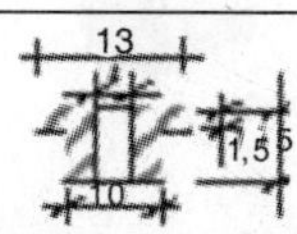
Powder	Sint B
Weight	3.60 gr

	Required Tolerance	**Statistical C_m Value Reached**
Radial run out	± 0.013 mm	1.96
Total Height	± 0.05 mm	2.61
Height flange	± 0.05 gr	2.77
Weight	± 0.08 gr	3.01

Note: A C_m value of 1.67 is generally accepted in production

$$C_m = \frac{\text{upper limit} - \text{lower limit}}{6\,S} \qquad S = \text{Standard deviation}$$

MECHANICAL ADAPTOR SYSTEMS

Standard adaptor single-platen version	Standard adaptor 2 lower platen

MECHANICAL STANDARD ADAPTOR - SINGLE PLATE

DESCRIPTION
- Standard adaptor for TPA presses in the range from 6 t... 70 t
- One upper and one lower level (upper punch, die)
- High supporting forces
- Usable on standard presses and on high-speed presses
- Usable for single-level and for stepped parts (de and core rod)

MECHANICAL STANDARD ADAPTOR - SINGLE PLATE

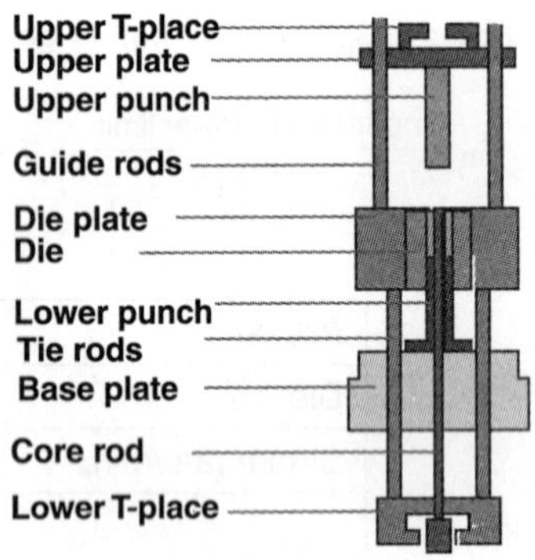

MECHANICAL STANDARD ADAPTOR - TWO PLATENS

DESCRIPTION
- Based on standard adaptor design
- One upper platen - two lower levels
- 1st lower platen: adjustable filling position fixed pressing position
- 2nd lower platen: adjustable filling position pressing position in the tool
- External hydraulic and PQC3 necessary

MECHANICAL STANDARD ADAPTOR - TWO LOWER PLATENS

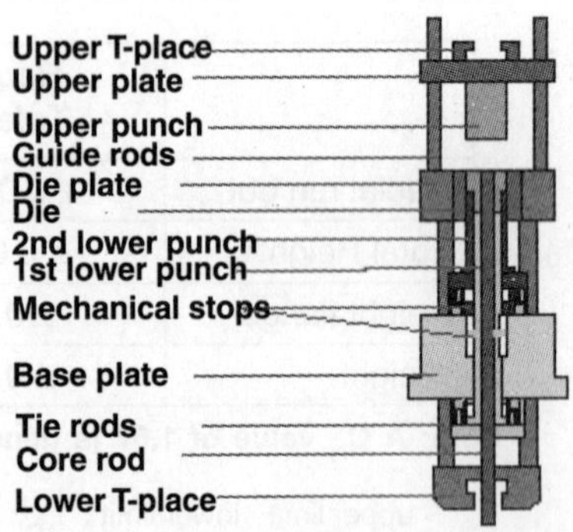

Fig. 9.7: Adaptors (die sets) used on high-speed presses

Naturally there are also some limitations in the use of high-speed presses:

- It makes no sense to press high parts at high stroke rates, filling is disastrous.
- Application of handling equipment has to be adapted.

But nevertheless stroke rates are increased by the factor 2–2,5 as compared with standard presses on compacts shown below.

Fig. 9.8: Typical parts for high-speed presses

High output rates make full automation advisable as 30 to 70 parts/minute or even twice as much -if double cavities are used can no longer be handled manually.

DORST delivers the necessary handling equipment tailor-made to the customer's requirements.

The benefits of high-speed presses can be summarized as follows:

- High productivity
- Economic efficiency
- Combination with automation concepts
- Operation 24 hours a day
- No manual handling, no handling faults
- Constant high quality

❍ ❍

10

Improve Overall Equipment Effectiveness on Powder Compacting Systems

Stefan Haltner, Alex Wehrli Peter Kunz and Markus Eismann,
OSTERWALDER AG, Switzerland

Abstract

Overall Equipment Effectiveness or OEE is basically the ratio of fully productive time versus planned production time. In practice, OEE is calculated as the product of the following three contributing factors:

Availability

Performance

Quality

This manuscript contains important information on where and how to improve overall equipment effectiveness. OEE Overall Equipment Effectiveness is a very important instrument to evaluate the equipment efficiency and especially to evaluate where the biggest loss of OEE occurs. If those factors and influences are understood, it is possible to work systematically against the losses of OEE.

10.1 CALCULATIONS

Overall Equipment Effectiveness is calculated as follows:

OEE = Availability * Performance * Quality

Availability:

Availability = Operating time/Planned production time

The planned production time does not take into account the planned shutdown time. If a factory is running only two shifts per day and five days a week, then this is considered planned production time. If in addition to that, the company has a two week planned shutdown, those two weeks are also deducted before we achieve the planned production time. So planned production time is equal to 100% OEE (refer to Table 10.1).

Table 10.1

Availability				
	Down time loss unexpected:			
		Equipment failure:		
		Machine:	142	Hours
		Tooling:	116	Hours
		Material shortage:	28	Hours
		Diverse down time:	122	Hours
		Total time loss unexpected:	408	Hours
	Down time loss expected:	Change over time:		
		Time per change over	96	Minutes
		Change overs per week	3	
		Weeks of production	45.14	Weeks
		Change over time total	216.69	Hours
		Total downtime	624.69	Hours
		Operating time	**3575.31**	**Hours**
		Availability:	**85.13**	**%**

Performance:

Performance = Nominal cycle time/(operating time/total pieces)

The operating time/total pieces equals the actual cycle time. So the performance of the equipment is finally Nominal cycle time/Actual cycle time.

Example: If the nominal cycle time is 5s and the actual cycle time is 6s, then the performance is 83.33% 5/6 = 0.8333 = 83.33%

Quality:

Quality = Good pieces produced/Total pieces produced

Example: If the total pieces produced are 1 000 000 and the good pieces produced are 994 000, then the calculation is 994 000/1 000 000 = 0.994 = 99.4%

OEE = Availability * Performance * Quality

OEE = 0.8513 * 0.8333 * 0.994 = 0.7051 = 70.51 %

10.2 IMPROVEMENT OF THE AVAILABILITY FACTORS

The availability of a powder compacting system contains different main factors.

One factor is simply the change from one order to the next order. So the tool change and eventually powder change has direct influence on the availability of the equipment. Another factor is unplanned shutdown of production due to equipment or tooling failure. Furthermore material shortage and eventual diverse factors like personnel shortage influence the availability of a press.

Availability Factor Tool Change

Tool changing time becomes more and more important today since companies are producing just in time and no longer in bigger lots. It is not an exception today, that companies are performing three or four tool changes per week on presses in the upper tonnage range (> 500T). On larger presses (> 250T) it is common to change the tooling outside of the press. The entire die-set is being exchanged within less than 20 minutes. Besides an immense time saving in tool change, there are several other very important advantages. You are able to detect tooling errors in advance while the press with the other die-set is still in full production. So we do not generate unwanted down time of the equipment. If the tooling is being changed inside the press and we detect tooling errors, the press may be down for one day or even more. With todays systems it is also possible to link multiple presses, so all die-sets can be used in each powder press (refer to Fig. 10.1).

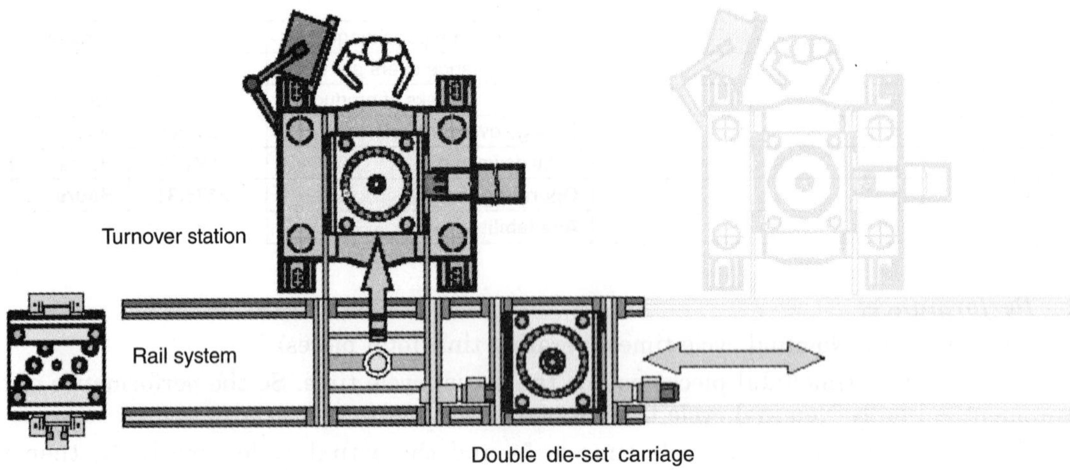

Fig. 10.1

Other advantages are the possibility of die-set revisions or other work on die-sets without taking the entire press system out of production. On the latest generation of smaller iron powder compacting systems (< 250T) in most cases the tool change is performed inside the press. Fast tool clamping systems for multi level tooling to minimize tool change times are available. Such systems allow changing times from last good part last good part to first good part in less than one hour Fig.10.2.

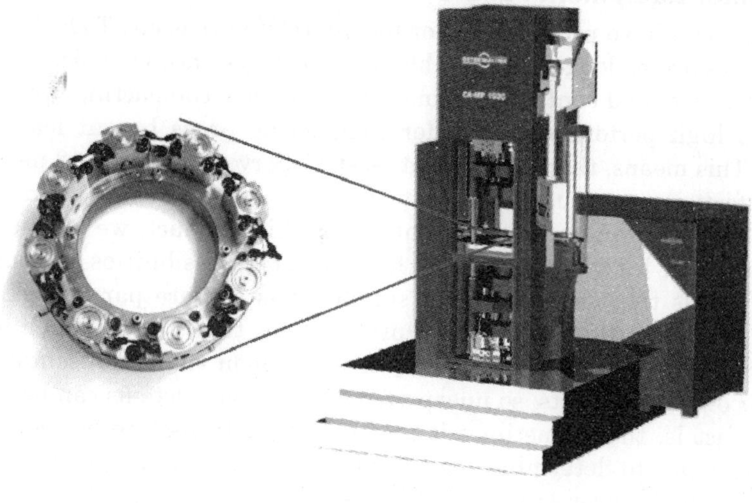

Fig. 10.2

On carbide presses for cutting inserts or for ceramic powder presses, there are various centreline clamping systems available. Such systems allow the change from one order to the next within 10 minutes. In the cutting insert and ceramic industry it is common to have several tool changes per day (refer to Fig. 10.3).

Fig. 10.3

Availability Factor Equipment Failure

Equipment failure can have press related or tooling related reasons. Today's high-end powder compacting systems are no longer comparable with older systems. On older mechanical presses there are less electrical and control equipment than on new compacting systems. It is today's standard that a high performance powder compacting press has at least 12 closed loop controlled axes. This means, that there are at least 12 Servo-valves and 12 measuring systems. No press manufacturer can guarantee that there are no failures of hydraulic or electrical components on high performance equipment. Since this is a fact, we need to evaluate what counter measures can be taken. Here we have several possibilities to counteract against those failures. One is that the equipment supplier has a spare parts stock which contains specifically all the parts that can shut a press down, then the supplier must be able to guarantee short delivery times. Additionally it is common that representatives around the world have their own spare parts, so unexpected delays from customs can be avoided. Another very important fact is, that sometimes it is not difficult to replace broken items or to ship replacement parts, but to determine the actual cause of the problem. One of the important services from a manufacturer is remote diagnostics via modem and/or a 24 hour helpline.

Availability Factor Material Shortage

Material shortage has direct influence to the availability of the compacting equipment. There are two main reasons for powder shortage. One reason is simply that there is not enough powder available in the factory, which is usually not the case. The second reason is, that some powder compacting presses do not have warning systems such as powder sensors. Those sensors have to be placed in a way, that the operator has enough time to change the big bag without stopping production. The sensors also have to be placed in a way, that there is always enough powder in the lower hopper to avoid weight fluctuation and finally to avoid bad parts.

Availability Factor Diverse Influences

There are many outside influences on the availability of a powder compacting system. Many of those influences cannot be predicted or manipulated. There may be naturally caused downtime through earthquakes or lightning, there may be personnel problems due to sickness or strike or simply general personnel shortage, and there may be things we simply can't predict. The aim of modern powder compacting systems is, that after a power outage or other outside influences, the press is back in full production as quickly as possible.

10.3 PERFORMANCE FACTORS

When the performance should be 100%, then the nominal stroke rate and the actual stroke have to be for all different press parts the same. There are different reasons why some of the movements during a cycle may have to be slower. For example, the filling time has to be increased due to bad weight consistency (*i.e.*, regrind powder, etc.), or the ejection speeds have to be reduced due to cracked parts (different tooling, etc.). Fact is also, that today's powder compacting systems are often not the limiting factor to the stroke rate, but rather furnace capacities, tooling heat up, filling time, etc., are the reasons for the lower nominal stroke rate. So in other words, often it is possible that when we need lower ejection speeds or more filling time, we can afterwards increase other speeds so the stroke rate remains at nominal rate.

10.4 QUALITY FACTORS

The quality is simply calculated as: good parts/total parts. Of course, some of the bad parts are being produced because of incorrect operation of the press. On the other hand, we have the powder press system which is responsible to make accurate and precise parts with outstanding CPK values. Modern powder compacting systems have several functions in order to improve the overall equipment effectiveness. First is the high accuracy and repeatability which is demanded, second is the user interface software of the press, which makes it possible that less incorrect operation of the press takes place. The user interface software makes it possible that the system controls the process and not the operator. Other functions are automatic surveillance systems like automatic filling height correction or automatic part height adjustment. Such highly developed systems allow the press to run more consistently and to produce less rejected parts. Through an intelligent user interface we are able to achieve three stroke optimization (refer to Fig. 10.4). On a known job it is possible to have good parts at the third pressed part.

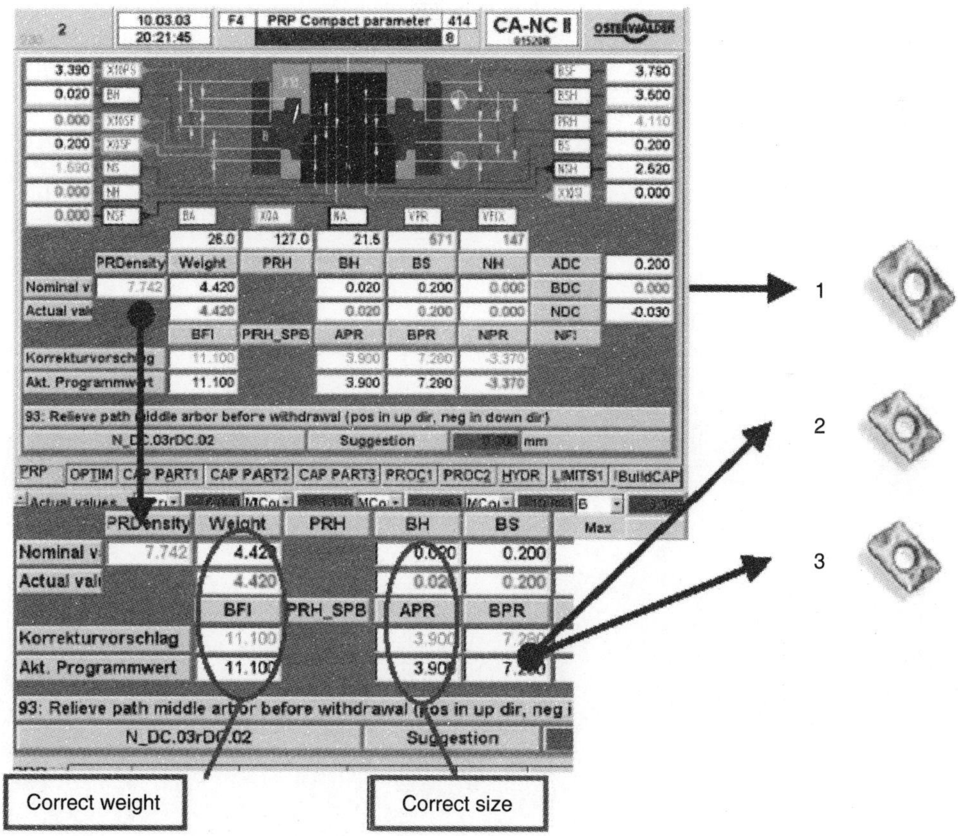

Fig. 10.4

10.5 CONCLUSIONS

If the equipment efficiency needs to be optimized, then it is essential that influencing factors are understood. So first step of OEE improvement is to figure out where the losses occur. Sometimes companies are concentrating on the dispensable factors instead to the often hidden influential factors. Consequently acquisition costs for quick-change tooling systems, die-set changing systems etc., have a very high return on investment because they increase the equipment effectiveness substantially.

Depending on a plant's production philosophy an increase of OEE on the powder compacting equipment may result in a significant productivity increase of the entire production line (press, sintering, sizing, etc.)

❍❍

11

Modern Hot-Pressing Technology Makes Sintering Quick and Easy

Schreyer,[1] Pascal, and Muralidhar, C.S.[2]
1. Dr. Fritsch Sondermaschinen GmbH, Fellbach, Germany 2. Diacore, Bangalore, India

Abstract

A fast-heating sintering process developed by Dr. Fritsch has been employed in many different industries for many years now. For instance, approx. 900 sintering presses based on this technology have been commissioned for the production of ceramics, diamond tools or friction pads.

11.1 INTRODUCTION

Hot pressing is a well known technique in the metal powder field. It is one of the sintering methods to consolidate cold pressed or loose powders. Uni-axial hot pressing is mainly used in the ceramic and diamond tool business, although there are also many other applications. Sintering temperatures of 650–2.400°C are normally used. Within hot pressing technology, three distinctly different types of heating can be found in use:

- Inductive heating
- Indirect resistance heating
- Dr. Fritsch "Direct resistance heating"

Inductive Heating Technology

In this process heat is produced within the mould when it is subjected to a high frequency electromagnetic field generated using induction coil coupled to an electronic generator. The mould is made out of graphite or steel, and pressure is applied by one or two cylinders onto the punches. The mould is positioned within the induction coil.

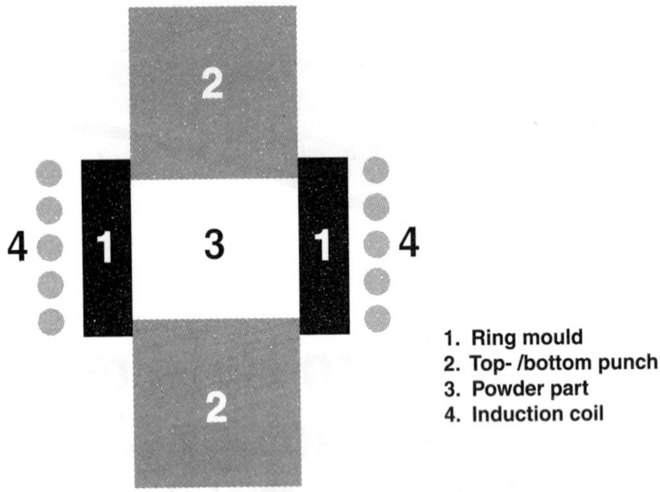

Fig. 11.1: Conventional inductive heating

The advantage here is that the pressure and the inductive power are completely independent. Even powders with a liquid phase are amenable to this process and low pressures are possible. Among the disadvantages are the expense of a high-frequency generator and the need for proper alignment. If the mould is placed off centre, the heat distribution is uneven. But the main disadvantage is the dependence of the process on good inductive coupling and heat conductivity of the mould. The magnetic field can penetrate the mould only 0.5 mm to 3 mm. From there on, the heat has to be "transported" into the mould by the thermal conductivity of the mould material. Uniform heating is much more difficult if the air gap between the mould and the inductive coil is not the same all along the mould profile. Another potential problem is heating rate. Too high a heat up rate will result in high temperature differences between the surface and core that can destroy the mould.

Indirect Resistance Heating

With indirect resistance heating technology, the mould is placed in a heating chamber. The chamber is heated by graphite heating elements. These elements are heated by electrical current. The heat is then transferred into the mould by convection. As the electrical energy heats the heating elements that then heat the mould in a secondary manner, the process is called indirect resistance heating.

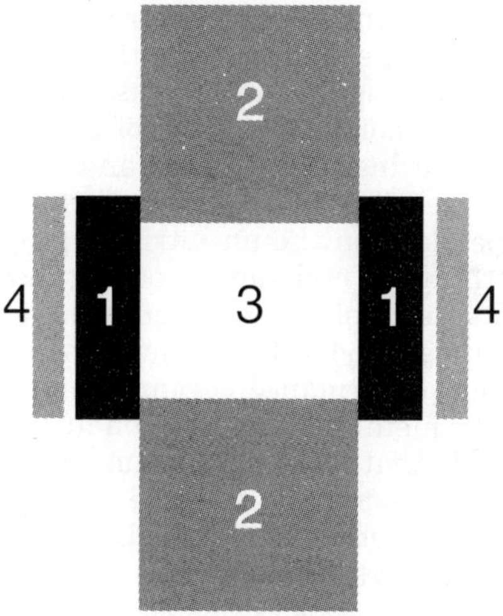

1. Ring mould
2. Top- /bottom punch
3. Powder part
4. Graphite heating element

Fig. 11.2: Indirect resistance heating

Advantages are high achievable temperatures, independent from the conductivity of the mould and independent from heat and pressure. Main disadvantage is the time that it takes to heat up the mould. It takes relatively long for heat transfer to take place from the furnace atmosphere to the mould surface and subsequently through out the cross-section of the mould.

11.2 DR. FRITSCH DIRECT HOT PRESSING

The basic idea of sintering with electric current going through the mould is quite old. Resistance heating of hard metal powders was patented by Tayler [1] as early as 1933. The compelling reason for shortening the cycle time then was to avoid grain growth and also save energy.

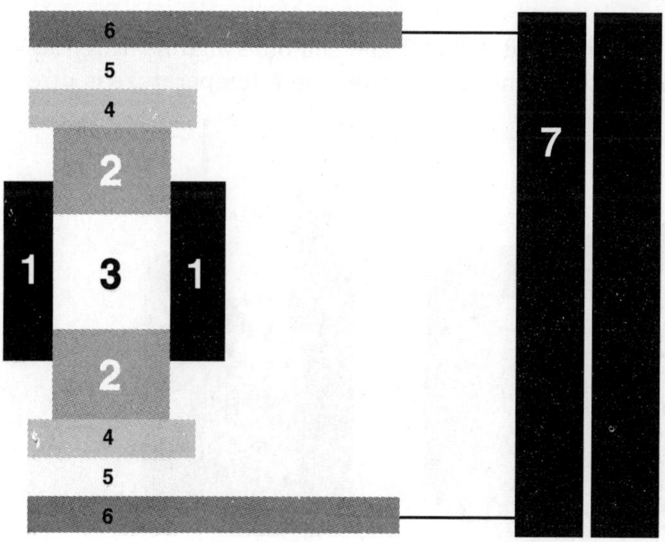

1. Ring mould
2. Top- /bottom punch
3. Powder part
4. Graphite electrode
5. Brass electrode
6. Copper plate
7. Transformer

Fig. 11.3: Dr. Fritsch DIRECT hot pressing

In direct resistance heating, the mould is directly connected to electrical power. The resistance of the mould and the powder part generates the heat directly in the mould. This results in very high heating rates. Additionally, this leads to significant increase in the sintering activity of fine metal powder aggregates which makes short cycle times of a few minutes possible. Further, this process lowers the threshold sintering temperature and pressure compared to that required in conventional sintering processes. The previous two methods as illustrated in Fig. 11.1 and Fig. 11.2 are both closely dependent on the an intrinsic property of the mould material, viz., its thermal conductivity. With direct resistance heating, however, the heat is generated where it is needed. Up to now this technology has been used mainly in the metal powder field with typical maximum temperatures of 1,400°C. Recently, the manufacture of such critical items as sputtering targets and high-performance ceramic components using B4C, TiDB2, SiAlON etc., has been achieved. Using metal powder, the conductivity of the mould is ideal for fast heating of the work-piece. Moulds that have a big diameter and relatively small height can be heated up very fast. The process is especially suitable for applications that need high heating rates, *e.g.* for materials that should not be kept at high temperatures too long or for processes that require fast heating rates for high productivity.

With the DIRECT hot pressing technology, materials can be sintered to their final density. The near net-shape precision achieved is very high and saves in many cases mechanical reworking of the high grade materials that are often difficult to process.

It can be used–amongst others–for the following materials and applications respectively:

- Innovative hard metals
- Multi-layer and composite materials
- High-performance ceramic materials
- Alloys, FGMs
- Special materials to be sintered with minor grain growth
- Sputter targets
- Friction pads

11.3 DR. FRITSCH DIRECT HOT PRESSES

Dr. Fritsch sintering presses have reached 2,400°C and combine high heating rates with high temperatures and precise temperature control. The workpiece temperature is directly

Fig. 11.4: Sintering press DSP 515 SA **Fig. 11.5:** Sintering press DSP 518 SA

measured by means of a pyroscope. One of the two standard machines with type DSP 515, has a maximum pressing power of 600 kN (60 tonnes) applied on a sintering surface of max. 220 × 220 mm, heating power is 150 kVA at 4.5 V to 6 V secondary voltage. The machine can operate at vacuum or under protective atmosphere; overpressure can also be realised.The machine has been successfully applied for metal and ceramic work pieces such as boron carbide at temperatures up to 2400°C. Process control and off-line programming via PC is also an option.

The sintering press DSP 518 has a maximum sintering surface of 350 × 350 mm. Mould heights of up to 320 mm are possible. The standard configuration is equipped with an 180 kVA transformer and a pressing power of 1.400 kN (140 tonnes).

Figure 11.6 shows the sintering chamber of a Dr. Fritsch sintering press. Every process parameter such as pressure, temperature, stroke and time are independently programmable.

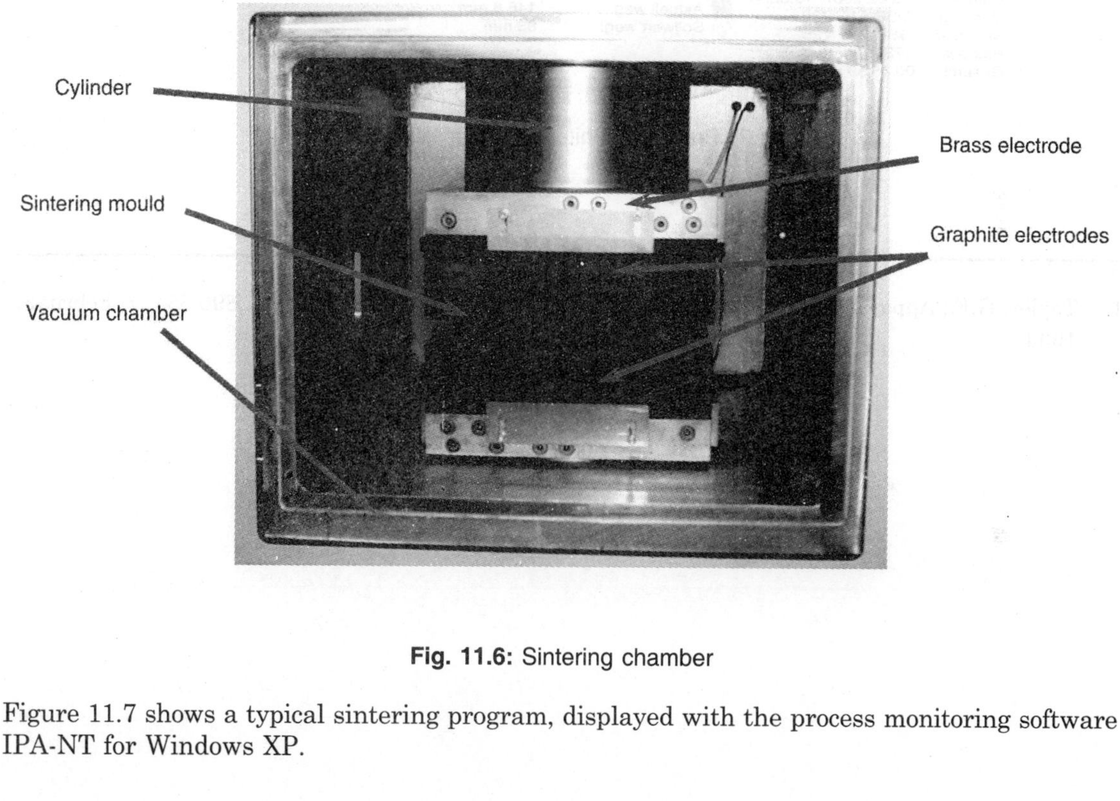

Fig. 11.6: Sintering chamber

Figure 11.7 shows a typical sintering program, displayed with the process monitoring software IPA-NT for Windows XP.

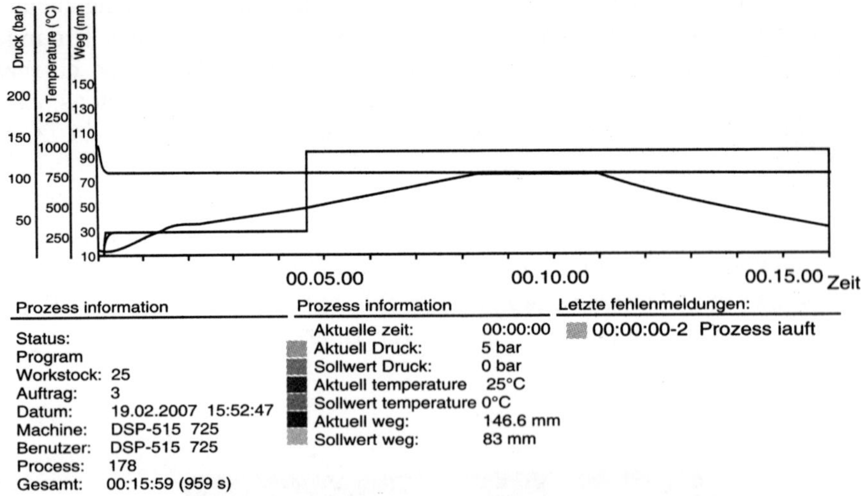

Fig. 11.7: Process monitoring under Windows XP

REFERENCES

1. Tayler, G.F.: Apparatus for Making Hard Metal Compositions, U.S. Patent 1,896,854, 7, February, 1933.

❍❍

12

Development of Automated Cleaning Technique of Top Plungers Inducted in Rotary Press Used for the Fabrication of PFBR Size Annular Fuel Pellets

S.K. Shrotriya, B.K. Shelke, Y.G. Nehete, A.K. Mishra and Mohd. Afzal
Advanced Fuel Fabrication Facility, Bhabha Atomic Research Centre, Tarapur, India

Abstract

Advanced fuel fabrication facility, BARC, Tarapur is presently fabricating uranium-plutonium mixed oxide fuel pellets for PFBR. The fuel pellets are annular and are of two different compositions as 21% PuO_2 and 28% PuO_2 and being fabricated vide powder metallurgical route involving cold compaction and sintering.

Rotary presses are used for the fabrication of mixed oxide annular fuel pellets due to high rate of production, simple design of tooling, easy way of tool changing and easy adoption of core rod feature. Existing design of tooling of Rotary press requires frequent clean up of top plungers due to settling of powder in the annular hole of top plungers during compaction. Delay in cleaning can also result in breakage of tooling apart from increase in the dose to extremities of personnel.

New design of tooling has been introduced to clean up the top plungers online instantaneously during the operation of rotary compaction press. It leads to increase in the productivity, reduces the spillage of feed material and also reduces the dose to extremities of personnel. The present paper describes the modification in the design of top plungers used in rotary press for online cleaning of top plungers during the fabrication of annular mixed oxide fuel pellets for PFBR.

12.1 INTRODUCTION

Advanced Fuel Fabrication Facility, BARC, Tarapur is manufacturing Mixed Oxide fuel for various thermal and fast reactors. The fabrication is carried out in α-leak tight glove boxes line to comply with safety requirements. It creates lot of constraints like space restriction, accessibility, maintainability and handling of material during the fabrication of plutonium based mixed oxide fuel. Hence, consideration of all aspects is required prior to selection of equipment [1].

Presently, mixed oxide annular fuel pellets for Prototype Fast Breeder Reactor is being fabricated on regular basis using depleted UO_2 and PuO_2 powders as raw materials by powder metallurgical route involving cold compaction and sintering as per flow sheet shown in Fig. 12.1. Mixed oxide annular fuels pellets for PFBR consists of 21% PuO_2 and 28% PuO_2 along with depleted UO_2 and are required to meet the physical and chemical specifications. Extensive experiments had been conducted with modified tooling design used in mechanical rotary compaction press for the optimization of process parameters to achieve desired specifications of fuel pellets.

12.2 MIXED OXIDE ANNULAR PELLETS FABRICATION FOR PFBR

Mechanical rotary compaction presses along with modified tooling were inducted for the compaction of ceramic grade MOX granules to fabricate annular fuel pellets for PFBR [2]. Toolings used in rotary press are made from Oil Hardened Non-Shrinking grade die steel. Core rods among top plungers, bottom plungers and dies are very prone to fail due to misalignment of die and plungers and incomplete die filling which causes eccentric loading. It is also subjected to wear during compaction over a time period. Hence, the core rods are being used having adequate toughness to overcome non-uniform stresses generated during compaction and also be capable of resisting ejection loads imposed. The tooling of rotary press used for the compaction of MOX powder to fabricate annular fuel pellets for PFBR are given in Fig. 12.2.

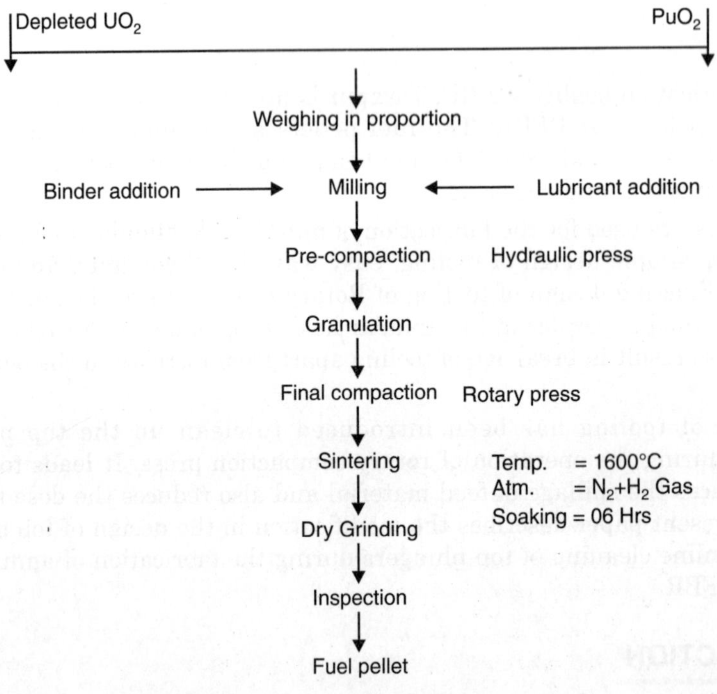

Fig. 12.1: Fuel pellet fabrication flow sheet

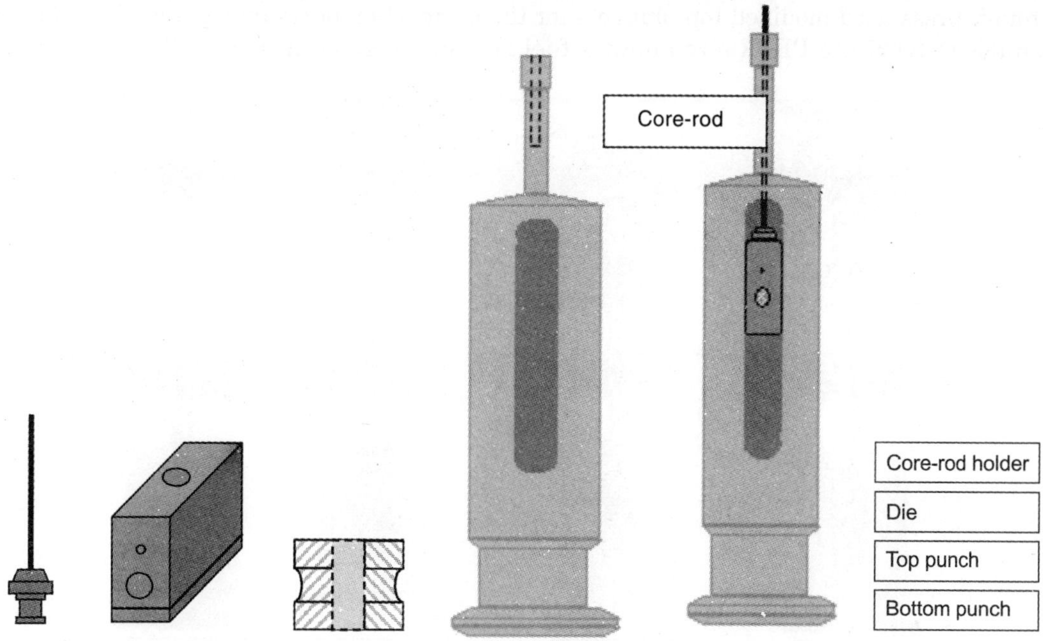

Fig. 12.2: Mixed oxide annular pellets fabrication for PFBR

12.3 MODIFICATION

In the existing tooling of rotary press used for the fabrication of PFBR annular fuel pellets, bottom plungers contain core rod whereas top plungers contain a central whole for the entry of core rod during compaction of powder. A little quantity of powder get settled in the hole of top plungers after few numbers of operations. Hence, top plungers have to be cleaned manually after certain number of operations. Modification in the existing tool design is initiated for online cleaning of top plungers that enhances the productivity and process efficiency.

Efforts are initiated to clean the top plungers instantaneously by providing a core rod in top plungers. The material and diameter of core rods used in both type of plungers are same. The position of core rods in both type of plungers are fixed in such a fashion that they should not interact with each other in any sequence of operation during compaction of powder. The core rod of bottom plungers is flushed with upper platform of the die whereas the core rod of top plungers is 3 to 4 mm projected from the top plungers. The up and down movement of top and bottom plungers are governed by separate cam arrangements. The modified top plungers used in rotary press move downwards during compaction sequence create vacant space in the central portion of top plungers for the easy entry of core rod of bottom plungers. Top plungers move upwards during ejection leaving vacant space in the central portion of top plungers. This action does the job of instantaneous cleaning of top plungers.

Extensive experiments were conducted using ceramic grade aluminum oxide granules with modified top plungers in rotary press to fabricate PFBR size annular pellets. The height of core rod and tooling size was decided on the basis of experiments. The compaction sequences of rotary press used modified top plungers for the compaction of ceramic grade aluminum oxide granules to fabricate PFBR size annular fuel pellets are given in Fig. 12.3.

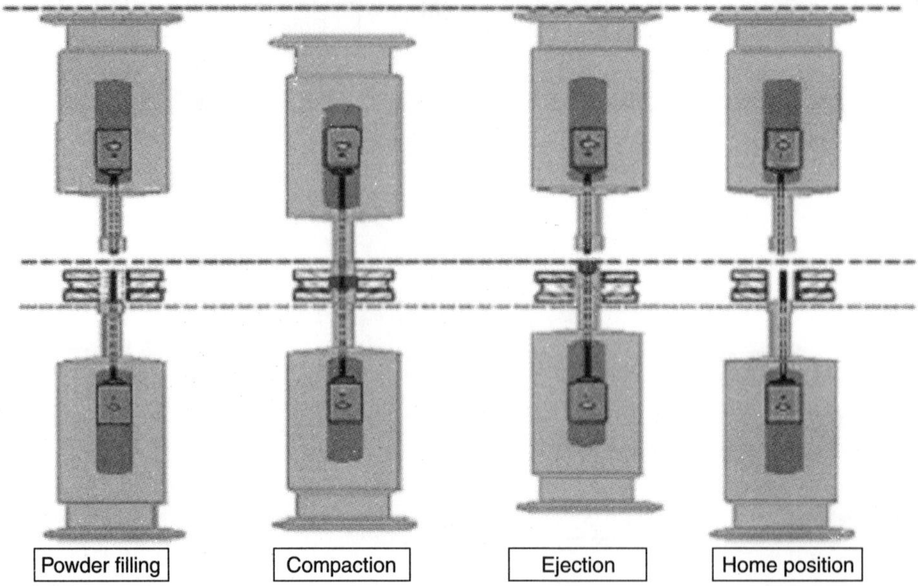

Fig. 12.3

With this modification cleaning of top plunger is carried out online instantaneously after each operation by providing a core rod in the top plunger. This core rod solves the purpose of top plunger cleaning while top plunger move upward during ejection sequence of rotary press. In this modification, the height of core rod used in top plunger is very crucial. If height of core rod used in top plunger is less then the required, it will come out from the central hole of top plunger during compaction sequence of rotary press resulting in misalignment during ejection sequence of rotary press. If height of core rod used in top plunger is more then the required, it may clash with core rod used in bottom plunger. To circumvent this type of scenario, a number of experiments were carried out to finalise the height of core rod used in top plunger.

Height of core rod may vary with different size of tooling sets. On the basis of these results, efforts are initiated to incorporate modified top plungers in rotary press used for the compaction of mixed oxide granules to fabricate annular fuel pellets for PFBR.

12.4 CONCLUSIONS

It has been demonstrated that the cleaning of top plunger is taking place instantaneously after each operation by providing a core rod in the top plunger. It enhances the productivity of the press and also reduces personnel exposure drastically. Moreover, this modification plays a key role in the reduction of spillage of valuable strategic nuclear material.

Acknowledgements

The authors wish to express their sincere thanks to Shri H.S. Kamath, Director, Nuclear Fuel Group and Shri J. P. Panakkal, Head, Advanced Fuel Fabrication Facility for their keen interest and valuable guidance extended to carrying out this work. The authors also wish to thank Shri P. S. Somayajulu, Ex. Superintendent, OFS of AFFF for his active cooperation in making this work possible. Special thanks are due to Shri V. K. Shumada and Shri S.G. Rane for their suggestions and help during carrying out this modification work.

REFERENCES

1. Arun kumar, Sudhir Mishra, A.K. Mishra, P.S. Somayajulu and H.S. Kamath, "Experience in Mixed Uranium Plutonium Fuel Fabrication", Transactions of Powder Metallurgy Association of India 24 [6] (1997), pp.38–42.
2. P.S. Somayajulu, Y.G. Nehete, B.K. Shelke, Arun Kumar and H.S. Kamath, "Development of annular pellet production techniques for PFBR fuel", PMAI, Hyderabad, 1999.

❍❍

13

The Magic of "Open Book Management (OBM)" ACuPowder International, LLC Case History

Edul Daver, ACu Powder International, LLC, U.S.A.

Abstract

Today's new economy–fast moving, entrepreneurial, fiercely competitive-demands a different way of thinking, with a new approach to motivation and action. OBM is a leadership concept, which unleashes magical results. It entails making employees understand the economics of business; think and act like owners; and have a stake in the success of the company. How does one do this?

With information. With business literacy. With empowerment. With a stake in the company. By making business, The Great Game. The article will discuss each element and how it is implemented at ACuPowder. It can transform a business and it has worked wonders at ACuPowder.

13.1 BACKGROUND ON ACuPOWDER INTERNATIONAL, LLC

ACuPowder is a manufacturer of Copper, Tin, Bronze, Brass and other Speciality Metal Powders. (Fig. 13.1) The major manufacturing process includes Atomization (Fig. 13.2) Metallurgical Reduction and Comminution. Its main plant is in Union, NJ, USA.(Fig. 13.3) with a satellite operation in Greenback, TN. 110 employees generate a total annual revenue of about $75 million. ACuPowder's customers utilize the powders to fabricate porous self-lubricating bearings, filters, friction components, diamond tools, brazing pastes, heat sinks and artistic creations. (Fig. 13.4). These components are used in automobiles, aircrafts, home appliances, lawn and garden, hand tools and computers.

ACuPowder was formed in 1995 as a result of a Management Leveraged Buy-Out of Alcan Metal Powders. The financing was raised with an "Asset Based" Financing structure. However, the deal was extremely leveraged with a Debt to Equity Ratio of 60:1. This required a new management strategy with effective motivation to jump-start the new company, which had very shallow pockets. An entrepreneurial operating philosophy was implemented with a generous incentive plan.

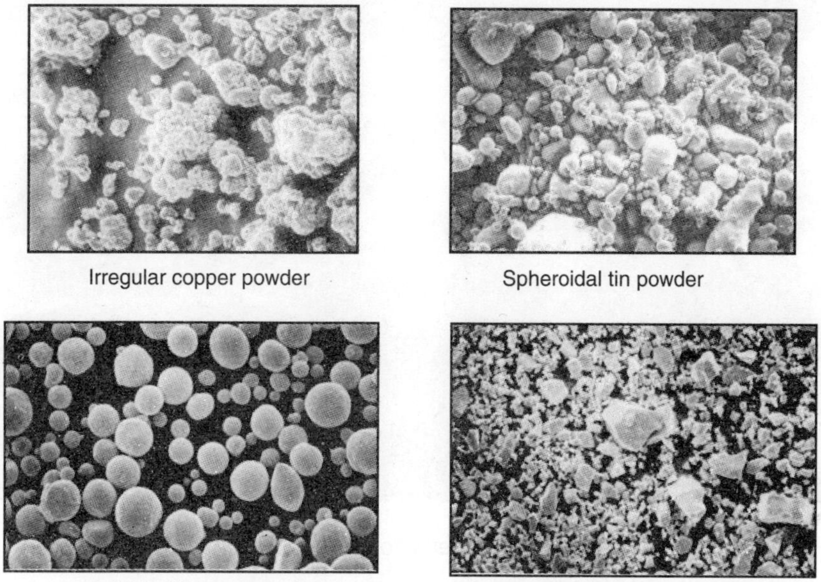

Irregular copper powder | Spheroidal tin powder

Spherical bronze powder | Angular manganese powder

Fig. 13.1: Metal powders manufactured by ACuPowder International, LLC

Fig. 13.2: Atomization process at ACuPowder International, LLC

Fig. 13.3: ACuPowder International, LLC, Union, NJ, USA

Fig. 13.4: Bearings, structural parts and filters made from ACuPowder's Powders

The formulated game plan was based on experience and intuition. It worked like magic and the infant company was increasing market share and its financial performance was much better then attained by the predecessor company. Within a few years, we came across a formal management concept called, "Open Book Management", which was remarkably similar to what we were practicing. Some missing elements were added to provide another boost. The infant company was performing remarkably well, well enough to purchase one of our competitors in 2000.

*Open Book Management: The Coming Business Revolution by John Case, Harper Business. 1994

13.2 INITIAL ASSET BASED FINANCING STRUCTURE

A $12,000,000 financing line was obtained based on the following formula:

Equity	...	$200,000
Land & Buildings	...	80% of Market Value
Equipment	...	80% of Liquidation Value
Receivables	...	85% of Eligible A/R
Inventory	...	60% of Eligible Inventory
Air Ball	...	$ Balance based on confidence in management and personal guarantee

The result was an extremely leveraged Debt/Equity Ratio (60:1), with no room for operating errors.

13.3 OPERATING AND MOTIVATIONAL PHILOSOPHY

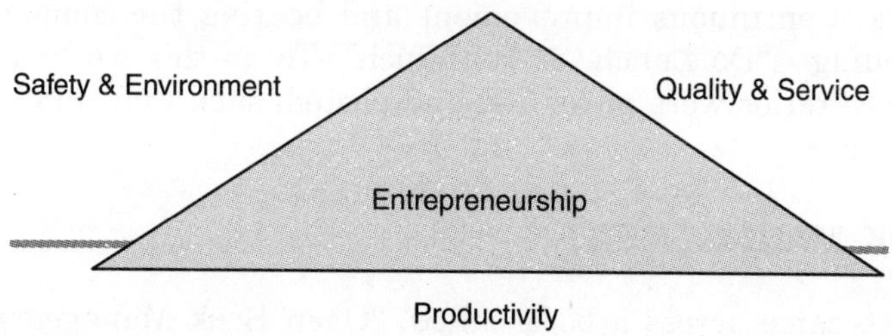

Entrepreneurship was defined as the pursuit of opportunity beyond resources currently available. A highly leveraged financing line; a flat management structure with focused objectives, skunk projects, swift decision making and utilization of established networks are some of the entrepreneurial tools which were constantly utilized. Productivity is basically the ratio for measuring output VS input-increase the output and reduce the input. An individual's productivity is best predicted by the following formula:

Productivity

$$= (\text{Ability}) \times (\text{Motivation}) \pm \text{Environment}$$
$$= (\text{Aptitude} \times \text{Training}) \times (\text{Need} \times \text{Incentive}) + \text{Environment}$$

Looking at this formula our focus was on the controllable, namely, Training, Incentive and Environment. Of these factors we felt that 'Incentive' was the most

Motivation

We had to change gears from being a division of a multi-national company to a small private company. As such our goal was to change from being Corporate managers to Entrepreneurial Leaders. To do this, seven key employees were offered to become Equity Partners. All employees, including hourly Union employees were promised Job security without lay-offs unless the company's survival was at stake. In addition, an Employee Incentive Award (EIA) was implemented as a powerful motivational vehicle.

Employee Incentive Award (EIA)

This award comprised distribution of 20% of pretax profits to all employees. This would be over and above the existing Union Contract and all other bonus award plans. The formula for distribution consisted of:

25%...Distributed equally to all employees
25%...Distributed based on salary levels
25%...Distributed based on length of Service
25%...Distributed based on departmental group performance

The total payout is then readjusted for absences and missed time. Each quarter a 'Shop Huddle' is held at which time the company's performance and financial results are discussed and EIA winnings announced. The actual payout is 50% of winnings at the end of the 2nd quarter, 75% after the 3rd quarter and 100% after the 4th quarter.

The Game of Business

We explained work as a serious but fun experience. Teamwork was essential and each player was expected to play their position within overall rules, but flexibility and pinch-hits were critical to teamwork. Continuous improvement and beating the competition were critical elements for succeeding - "Do Lunch, or Be Lunch". To do this we had to keep score and countless performance ratios were measured and posted, such that numbers did the talking and Profits would follow.

13.4 OPEN BOOK MANAGEMENT

In the late 1990s we came across a book called, "Open Book Management" by John Case. The cover statement said, "Companies that practice OBM seem to have captured some sort of lightening in a bottle." So we read the book and guess what? Without knowing it, we were practicing OBM. However, one critical item was missing at ACuPowder. We had not provided formal business literacy education to our employees to enable them to understand the relationship between what they did and the bottom line. As such, we got a local college to provide a Business 101 Course to all employees–financial terms, income statement, etc. Motivation and Results improved further. It was like magic, so let's discuss further...What is OBM? How to implement it? Why does it work?

13.5 OBM–WHAT IS IT?

- OBM is a way of running a company that gets every employee focused on helping the business make money-short and long term.
- OBM enables every employee to develop business literacy and understand company's critical numbers and financials.
- OBM employees accept responsibility and are empowered to move these numbers in the right direction.
- OBM employees see themselves as partners in the business with a direct stake in the company's fortunes.

13.6 OBM–HOW TO IMPLEMENT IT?

- Provide information and communication. The basic premise is that messages need to be repeated many times before it is understood. The first time it is 'heard', the second time it is 'listened' and third time perhaps it is 'understood'. The information disseminated includes: A clear vision with strategy and tactics. Annual and/or shorter term objectives, its measurement and visual charting. Results and Financial Attainment.
- Provide Business Literacy Training Every employee must understand Business Basics. Employees need to clearly understand financial and/or other constraints. Purpose is to enable better decision making and buy-in with less waste.
- **Empower Employees**
  - Employees have a substantial 'say' in ones own work.
  - Work teams huddle to discuss problems and/or opportunities.
  - Experimentation is encouraged and related failures accepted as a learning experience.
  - Continuous Improvement is the name of the game.

- Have a stake in the Success of the Company
  - Celebrate Successes
  - Build enthusiasm and positive peer pressure/motivation.
  - Share in the financial success of the company *e.g.*, cash bonus or equity distribution.

13.7 OBM–WHY DOES IT WORK?

- Common goals outweigh parochial ones.
- It focuses attention, like a laser, on objectives.
- It facilitates innovation and continuous improvement.
- OBM helps managers lead rather than manage.
- Financial reward catalyzes motivation and peer pressure.

- Sources of mistrust and resentment almost vanish.
- Make each employee feel and behave as a valued partner and entrepreneur.
- "Nothing breeds success more than success itself".

13.8 IT WAS MAGIC

The results we attained during our first five years were unbelievable and even surprised us. The market share, for the NJ plant increased from 19% to 40% today. In 2000 we had sufficient retained earnings to facilitate the buy-out of a smaller competitor located in Tennessee. Today, ACuPowder Companies, have captured over 50% of the North American market share making it the dominant #1 non-ferrous metal powder company in North America. We are very proud of the fact that currently we are able to export successfully to emerging countries, such as, India, Korea, Taiwan and yes even China. The healthy profits has resulted in EIA payouts to employees which have exceeded $10,000/yr/employee.

13.9 HURDLES AND CHALLENGES

It should be clarified and emphasized that results don't just happen and that management has to maintain focus and lead by example. A sound vision, good planning and superior innovative execution are prerequisites, which are catalyzed by Open Book Management Concepts. The biggest challenge is to keep OBM "Center Stage" and prevent it from becoming an entitlement.

The total 'kitty' which comprised 20% of pretax profits remained unchanged and was treated as being "untouchable". We strongly felt that tampering with the base amount would destroy our credibility and be counter productive. However, the Also, all changes, however small, must be done proactively so that employees are not surprised and can actively participate towards the attainment.

'Yesterday' has been very successful. However, we constantly have to remind ourselves that past performance is no guarantee of future success and that hubris is dangerous and can become lethal.

'Today' is a new challenge and we have to accept it and work together to "earn our wings everyday".

'Tomorrow' brings new business challenges and we have to pursue the related opportunities beyond the resources currently available.

13.10 CONCLUSIONS

OBM is a powerful concept and tool, which has contributed to ACuPowder's success. We expect that you can utilize some of its concepts in almost any business environment and innovatively fine-tune it to meet your needs. We hope it will be "magic" to you too. Good Luck.

❍❍

14

Studies on the Hydroxide Preparation, Calcination and Related Characterization in the Powder Metallurgy of Hafnium

Raghu C Reddy, M R P Reddy, R Govindaiah, Angad Choudhary, Arbind Kumar, and T L Prakash*
Centre for Materials for Electronics Technology, Hyderabad, India

Abstract

Hafnium is a refractive metal known since 1923 finding applications in the fields like space vehicles, modern aero engines, nuclear reactors, next generation electronic gadgets, etc. The world resources of hafnium contained in zircon ore is estimated to be over 1 million tons. Owing to the difficulties in separating Hafnium from zirconium due to their chemical similarity the production of their respective pure entities is quite challenging. The metallurgy of hafnium is highly complex and versatile. It involves a number of processes used in the conventional metallurgy starting from hydrometallurgy to pyrometallurgy using sophisticated vacuum and allied techniques. Hafnium in the form of hydroxide is obtained from zircon sand by solvent extraction technique. Conversion of hafnium hydroxide to its metal is brought about by a reduction step involving carbo-chlorination stage. Powder characteristics of hafnium oxide have to be optimum to efficiently convert it to its respective chloride. The fineness and morphology of the oxide depends on its formation conditions during hydroxide conversion and subsequent heat treatment. The oxide should also be devoid of in-process impurities, like phosphorous and nitrogen. The chlorine efficiency and product quality during carbo-chlorination of hafnium oxide powder depends on a number of parameters. Precise control of powder characteristics and subsequent powder processing to prepare the feed for chlorination are of paramount importance. This paper encompasses the studies conducted to sensitize the calcinations parameters for hafnium oxide powder suitable for chlorination process. The effect of the type of alkali and its concentration on the quality of hydroxide formed is also reported. Results of various characterization tools used for the powder like, laser particle size analysis, Fischer sub-sieve analysis, the BET surface area, ICP-OES, O-N analyzer, optical microscopy and SEM presented.

14.1 INTRODUCTION

Hafnium is an important metal finding applications in high temperature alloys. The abundance of hafnium is less from the known sources in the earth's crust and many times it is very difficult to produce its pure form. Apart from high temperature alloys, hafnium finds applications in other specialized fields like, atomic energy, space technologies and electronics. It is used in control rods of nuclear reactors and as an alloying element in high temperature alloys. Its use in electronics as the next generation gate material and as a nuclear isomer which can be used for energy storage applications are being investigated extensively world over.

Hafnium and zirconium have nearly identical chemistry, which makes the two difficult to separate [1, 2]. Liquid-liquid extraction processes also known as solvent extraction process is used extensively for the production of hafnium. About half of all hafnium metal manufactured is produced as a by-product of zirconium refining. The hafnium from the hydrometallurgy is then converted into its oxide; hafnium(IV) chloride. In the chlorination process, petroleum-derived coke along with hafnium oxide is made to react with chlorine gas in a fluidized/packed bed reactor at 1100 °C. The volatile chloride of hafnium formed in the process is collected in a condenser at lower temperatures. In a separate reactor, the $HfCl_4$ is reduced by liquid magnesium (15–20% excess) at 1100°C in a stainless steel retort to ensure complete reduction in the Kroll process as per the reaction, $HfCl_4 + 2Mg\ (1100\ °C) \rightarrow 2MgCl_2 + Hf$.

In the multi-step process involving conversion of hafnium hydroxide to its metal, the powder characteristics of intermediate hafnium compounds have to be optimum for their efficient reaction. Precise control of powder characteristics and subsequent powder processing to prepare feed for chlorination process is a critical step in the processing of hafnium. The fineness and morphology of the oxide depends on its formation conditions during hydroxide conversion and subsequent heat treatment [3,4]. The oxide should also be devoid of in-process impurities, like phosphorous and nitrogen. The chlorine efficiency and product quality during carbo-chlorination of hafnium oxide powder depends on a number of parameters. This paper encompasses the studies conducted to sensitize the calcinations parameters for hafnium oxide powder suitable for chlorination process.

14.2 EXPERIMENTAL

The tailings from the zirconium extraction called the zirconium scrub raffinate contain hafnium in very low concentrations; few PPM by weight. The extraction of hafnium is carried out in slurry extraction system shown in Fig. 14.1. The solvent extraction process is carried out in two steps to achieve 99% hafnium with respect to zirconium. The chemical analysis was carried out using the Inductively Coupled Optical Emission Spectroscopy (ICP-OES) through out the extraction process in achieving the required purity. As shown in the flow chart at Fig. 14.2, initially, hafnium is enriched by removal of zirconium from the solutions. Tributyl Phosphate (TBP) is used as a solvent diluted with kerosene in the ratio 1:2. The aqueous feed is prepared by using nitric acid of normality 4 to 4.5 N.

Fig. 14.1: Slurry extraction system used for the solvent extraction of hafnium

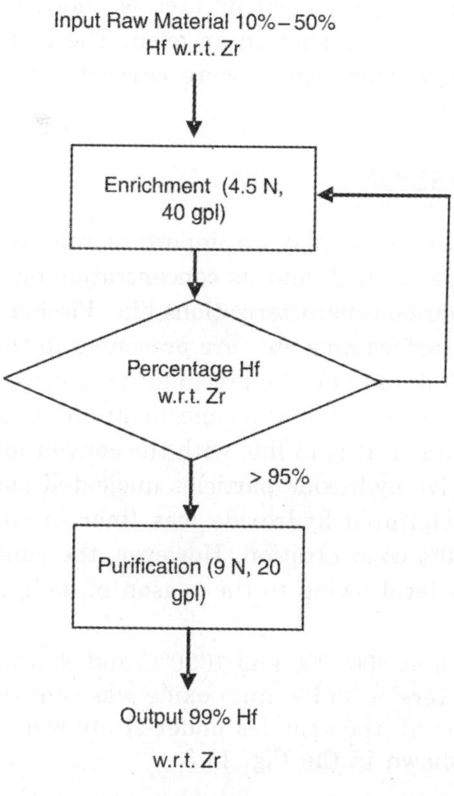

Fig. 14.2: Flow sheet hafnium processing

In the slurry extraction system, which is a counter current type pneumatically operated solvent extraction system; zirconium from the aqueous feed gets extracted to the organic solvent, bringing about enrichment of hafnium. Typically, solvent extraction is carried out in seven stages constituting one cycle. Depending on the enrichment required, these cycles are repeated. When the purity of hafnium reaches more than 95%, it is taken through a purification cycle, where co-extraction (9 N) and selective stripping (3 N) is carried out to get pure hafnium in nitrate solution.

The nitrate strip solutions containing hafnium of purity 99% with respect to zirconium was used in this study as the staring material. 20 litres of nitrate strip solutions containing around 6 gram per litre of hafnium was taken in each batch to study the effect of processing on the oxide produced. The nitrate solutions when neutralized by a base precipitate hafnium in the form of its hydroxide. The hydroxide is heat treated in the calcination process to get hafnium oxide. The experiments were designed to study the effect of precipitating media and calcination temperatures on the resultant powder morphology of hafnium oxide. The effect of alkali concentration was studied by making use of sodium hydroxide at 10 and 20% concentrations by weight. Higher concentration of alkali greater than 20% is known to partially de-hydrolyze the hafnium hydroxide formed due to large amount of heat of reaction. Ammonia gas at a constant flow rate of 5 litre per minute was used as the other alkali neutralizer to produce hafnium hydroxide. After neutralizing the nitric acid solutions containing hafnium to 7 pH, the hydroxide precipitated is allowed to settle, the supernent decanted and the precipitate is filtered, dried in an oven. The hydroxide is then taken for calcinations heat treatment. The calcinations were carried out in a muffle furnace for durations of 1 hour in each case.

14.3 RESULTS AND DISCUSSION

Powder characteristics of hafnium oxide play an important role in its optimum conversion to chloride. The effect of the type of alkali and its concentration on the quality of hydroxide formed was studied. Results of various characterizations like, Fischer sub-sieve analysis; laser particle size analysis, the BET surface area etc., are presented in the Table 14.1. It is found that, the concentration of the alkali used for the precipitation process affects the final particle size of the hydroxide obtained. The higher the concentration of the alkali, finer was the particle size obtained and vise-versa. It is in line with the conventional wisdom; in that, the leaner reactant concentration give hydroxide particles nucleated more time to grow. It was also found that, ammonia precipitated hydroxide was finer in comparison to the NaOH precipitated hydroxide up to 20% concentration. However, the concentration of this alkali cannot be increased above this level owing to the reason of de-hydrolysation of hydroxide rendering it less reactive.

Calcinations were conducted at 800, 900 and 1000°C and characterization of the powder was carried out. The phase conversion to hafnium oxide was confirmed by the powder XRD analysis carried on the samples; all the samples under study were completely converted to phase pure hafnium oxide as shown in the Fig. 14.3.

Table 14.1: Tabulated results

Sample	Precipitation Medium	Calcination Temperature Time (Degree C)	Laser Particle Size (micron)	Bulk Density (g/cc)	FSS (micron)	BET Surface Area (square m/g)
Hf-Hydroxide	10% NaOH	–	36	1.53	0.70	3.7064
Hf-Hydroxide	20% NaOH	–	29	1.42	0.60	3.2865
Hf-Hydroxide	Ammonia	–	22	1.63	0.90	3.3399
Hf- Oxide	10% NaOH	800	33	1.76	0.98	4.6935
Hf- Oxide	Ammonia gas	800	18	2.78	0.95	27.8899
Hf- Oxide	Ammonia gas	900	22	3.21	1.38	16.3482
Hf- Oxide	Ammonia gas	1000	23	3.20	1.40	11.6812

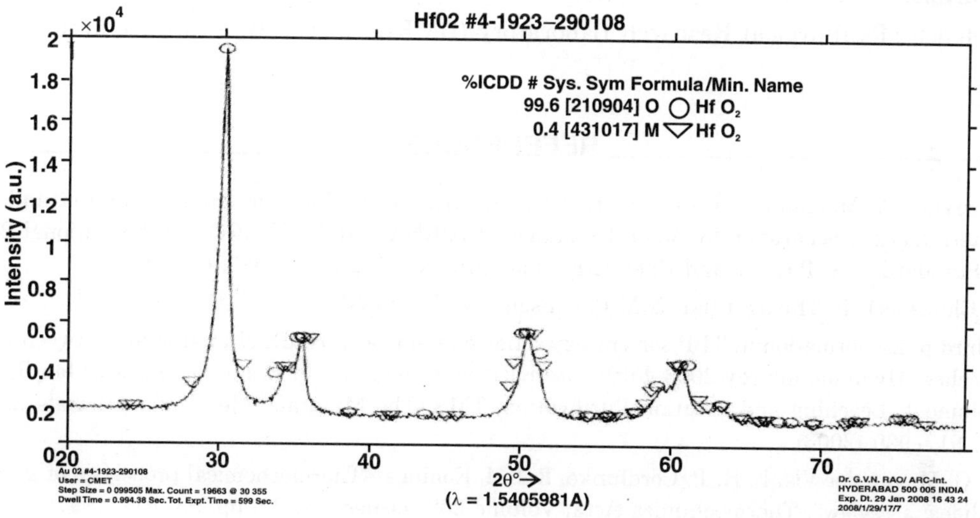

Fig. 14.3: X-Ray diffraction pattern of the hafnium oxide

It was found that, samples calcined at 800°C for 1 hour duration were having the maximum BET surface area. Increase in calcinations temperature resulted in progressive decrease in surface area with the temperature increase. This is attributed to sintering and agglomeration of oxide particles due to higher calcinations temperatures leading to reduction in surface area. The other characteristics determined, *viz.*, bulk density and FSS particle size did not show any fixed trend with respect to calcinations temperature and BET surface area. However it will be very useful if a correlation can be found between these characteristic and the surface area since, the instruments for evaluating them are less sophisticated and are generally easily available unlike BET surface area and laser particle size analysis. It was also noticed that, hafnium oxide calcined from NaOH precipitated hydroxide did not give the desired high BET surface area with only marginal increase in surface area from its hydroxide.

14.4 CONCLUSIONS

- The BET surface area of hafnium oxide depends on the manufacturing route followed in producing it.
- Ammonia gas when used for the precipitation of hafnium hydroxide is found to render better surface area to its oxide in comparison to sodium hydroxide.
- A calcination temperature of 800°C when applied for duration of one hour is found to give the optimum surface area hafnium oxide.

Acknowledgements

1. Vikram Sarabai Space Centre (VSSC), Trivendrum for the funding of the project under which this work was carried out.
2. Nuclear Fuel Complex (NFC), Hyderabad for the raw material.
3. National Centre for Compositional Characterization of Materials (NCCCM) for the chemical analysis.
4. Defence Metallurgical Research Laboratory (DMRL) for the BET surface area analysis

REFERENCES

1. Lawrence A. Machlan and John L. Hague, "Separation of Hafnium from Zirconium and Their Determination: Separation by Anion-Exchange", JOURNAL OF RESEARCH of the National Bureau of Standards—A. Physics and Chemistry, Vol. 66A, No. 6, pp. 517–520, (1962).
2. R. Chiarizia1, P. Thiyagarajan 2, M.P. Jensen1, M. Borkowski1, and K.C. Littrell.
   "Third phase formation in TBP solvent extraction Systems as a result of interaction between reverse micelles" Hydrometallurgy 2003–Fifth International Conference in Honor of Professor Ian Ritchie – Volume 1: Leaching and Solution Purification; TMS (The Minerals, Metals & Materials Society), pp. 917–920 (2003).
3. M. G. M. van der Vis, E. H. P. Cordfunke, R.J.M. Konings, "Thermochemical properties of zirconium halides: a review", Thermochimica Acta, Volume 302, Issues 1-2, 22, pp. 93–108 (1997).
4. Robert G. Jones, "Halogen adsorption on solid surfaces", Progress in Surface Science, Volume 27, Issues 1-2, pp. 25–160 (1988).

○○

15

Preparation of High Purity Sub-Micron Spheroidal Zirconia Powder from Impure Zirconium Salt through Glycerol Route

Amit Sinha, P.K. Sinha and B.P. Sharma
Bhabha Atomic Research Centre, Navi Mumbai, India

Abstract

Using a simple energy efficient process, high purity fine zirconia powder with uniform particle morphology was prepared starting from impure zirconyl chloride. The process involves hydrolysis of zirconium in glycerol media and subsequent removal of impurity elements by controlling the process parameters. Glycerol, in this process, acts as a solvent as well as a reducing agent. The zirconia powders prepared through this process were characterized by X-ray diffraction (XRD), scanning electron microscopy (SEM), BET surface area analysis, thermo-gravimetric analysis and chemical analysis. The zirconia powder produced through this process was having a purity greater than 99.9 %.

15.1 INTRODUCTION

Pure zirconia (ZrO_2) is one of the major constituents of ceramic colors and an important component of lead zirconia titanate based electronic ceramics. Pure zirconia is monoclinic (m) at room temperature. During heating, m-ZrO_2•t-ZrO_2 (tetragonal) and t-ZrO_2•cZrO_2 (cubic) phase transformations take place at 1170 and 2370 °C respectively. The cubic phase is stable up to the melting point (2680 °C) of zirconia. The t-ZrO_2•m-ZrO_2 phase transformation is accompanied by a large volume change (3–5%) leading to disintegration of the ceramic body during cooling from the sintering temperature and subsequent application. However, addition of aliovalent oxide additives *i.e.*, CaO, MgO, or Y2O3 into the zirconia stabilizes the high temperature cubic phase through the formation of vacancy at oxygen sublattice. The cubic zirconia is used in grinding media and engineering ceramics due to its increased hardness and high thermal shock resistance. It is also used in applications such as oxygen sensors and solid oxide fuel cells due to its high oxygen ion conductivity. Preparation of pure zirconia is usually carried out starting from zircon ($ZrSiO_4$). In this process, the first step is to convert zircon to zirconyl chloride. It can be done through the following chemical reactions:

$ZrSiO_4 + 4NaOH , Na_2ZrO_3 + Na_2SiO_3 + 2H_2O$ (1)

$Na_2ZrO_3 + HCl , ZrOCl_2. 8H_2O$ (2)

For the preparation of zirconia powder from zirconyl oxychloride, there are basically two methods, *e.g.*, thermal decomposition and precipitation. In thermal decomposition, once $ZrOCl_2. 8H_2O$ is heated to 200°C it starts dehydrating and becomes anhydrous $ZrOCl_2$ which on subsequent heating forms zirconia. This method is associated with low production cost. The limitations of this process include large impurity content in powder as well as difficulty in producing powder with fine particle size. Precipitation method, on the other hand, uses chemical reaction to obtain zirconium hydroxide as an intermediate which forms zirconia powder on subsequent calcination.

The application of advanced ceramics in structural and electrochemical applications calls for a fine, uniform microstructure along with very high sintered density. This in turn calls for oxide powders with outstanding characteristics in terms of particle morphology, composition and purity. One of the promising routes for synthesis of high-quality ultrafine ceramic powders is wet-chemical method [1]. For the preparation of ceramic powders through precipitation from a liquid starting salt, solvents such as polyols offer two advantages: the high dielectric constants of polyols facilitates dissolution of inorganic compounds and their high boiling point offer a wide operating temperature range (room temperature to boiling point) that permits hydrous reaction that cannot be possible in aqueous solution under atmospheric pressure. These potentials of polyols have been greatly exploited for preparation of metal powders with controlled morphology [2, 3].

There have been a few reports on the preparation of ceramic powders using polyol medium. Matijevis and Climas prepared uniform colloidal iron (III) oxide in water–ethylene glycol solution [4]. Typical processes that utilize inorganic salts as precursors for solution-synthesis of oxide ceramic powders include forced hydrolysis or other hydrothermal processing, that usually calls for relatively high temperatures to induce hydrolytic reactions, condensation, and the subsequent formation of oxide particles. As kinetics of the forced hydrolysis process is relatively slow, one of the strategies to enhance the particle formation kinetics is to add a low dielectric medium to the aqueous solutions of inorganic salts. Controlling the dielectric property of the mixed alcohol–water solvent influences not only the solubility of the dissolved solute but also the colloidal interaction between solid particles.

In the present article, we report the preparation of high purity nanocrystalline spheroidal zirconia powder from impure zirconyl chloride salt using a polyol process. The same process has also been utilized for preparation of yttria stabilized zirconia powder.

15.2 EXPERIMENTAL

Impure zirconyl chloride ($ZrOCl_2$) containing 1 wt % iron impurity (Indian Rare Earth, India) was used as starting material for the preparation of zirconia powder. A predetermined amount of starting salt was mixed with glycerol (AR grade). The mixed solution was heated, while stirring, on a laboratory hot plate. The reaction temperature was measured using a thermometer submerged into the solution through a glass port. After the precipitation of precursor powder, the suspension obtained was naturally cooled to room temperature and then diluted with ethanol. The Zr-containing precipitates were recovered from the suspension by repeated centrifugation followed by washing with ethanol. The powder obtained was dried under vacuum at 120°C for 1 h. Decomposition behaviour of Zr^{+4} precursor was studied by thermo-gravimetric analyzer up to a temperature of 800°C under flowing air atmosphere at a heating rate of 10°C/min. Based on the thermal analysis data, the precursor was calcined at 650°C for 2h. The phase analysis of the dried precipitate as well as the calcined powder was carried out by X-ray diffraction using CuK. radiation at a scanning rate of 0.05 °/sec. The specific surface area of the powder was measured by conventional nitrogen BET surface area analyzer. The morphology of the calcined powder was studied by scanning electron microscope.

For preparation of 8 mol% yttria stabilized zirconia (YSZ) powder, $Y(NO_3)3$ was used as starting material along with $ZrOCl_2$. The precursor powder obtained from the process was calcined at 700°C for 1 h to form the desired phase. The phase analysis of the calcined powder was carried out using room temperature. To study the evolution of phases as a function of temperature and also to study the crystal growth behaviour of the calcined powder at different temperatures, high temperature XRD study was conducted on calcined powders.

15.3 RESULT AND DISCUSSION

15.3.1 Preparation of Pure Zirconia Powder

The flow sheet of preparation of zirconia powder through glycerol route is shown in Fig. 15.1. When a zirconyl oxychloride solution with an alcohol-water mixture as a solvent is heated, the dielectric constant of the solution decreases significantly with increase in temperature leading to super-saturation of salt solution and subsequent precipitation [5]. In the present investigation, it was observed that Zr^{+4} started to hydrolyze and precipitate out of the solution at 140°C.

The TG curve for zirconium precursor, as shown in Fig. 15.2, shows that decomposition is complete below 550°C. Based on the thermal analysis data, the precursor was calcined at 650°C for 2 h. The XRD pattern of the calcined precursor is shown in Fig. 15.3. The pattern exhibits the diffraction peaks corresponding to of tetragonal and monoclinic phases of ZrO_2.

The SEM photomicrograph of the zirconia powder produced through this route is shown in Fig. 15.4. The micrograph reveals that the powder is spheroidal having narrow size distribution. The crystallite size calculated from X-ray line broadening using the Scherreer formula was 12 nm. The BET specific surface area of the powder is 48 m^2/gm which corresponds to a crystallite size of 20 nm. The chemical analysis of the powder indicated that it contained less than 100 ppm of iron and purity of the zirconia powder was greater than 99.9%.

The above results suggest that powder produced through this route exhibit a very high purity and a good crystallinity. The particles have a regular and polyhedric shape *i.e.*, their growth is isotropic.

15.3.2 Preparation of Yttria Stabilized Zirconia

The XRD pattern of YSZ powder prepared through glycerol route which was calcined at 700°C for 1 h, is shown in Fig. 15.5. The pattern exhibits all the peaks corresponding to cubic polymorph of zirconia. The crystallite size of the calcined powder was calculated from the line broadening of the strongest XRD reflection. To calculate the crystallite size of the powder as a function of temperature, the XRD patterns were recorded in the temperature range of 700 to 1000°C. The average crystallite size of the YSZ powder heated to different temperatures was determined from the line broadening observed for the peak corresponding to the (101) reflection using the Scherrer's formula. The YSZ powder was found to be nanocrystalline having a crystallite size of 6.8 nm at 700°C and 10 nm at 1000°C. It was observed that the calcination temperature has greater influence on final crystallite size than calcination time.

15.4 CONCLUSIONS

A simple energy efficient process has been described for preparation of zirconia powder starting from impure $ZrOCl_2$ as a precursor for zirconia. It essentially consists of dissolving the zirconia precursor in glycerol followed by hydrolysis and precipitation from the solution. The process yields nano-crystallized zirconia powder having a purity greater than 99.9%.The XRD analysis of the powder calcined at 650°C indicates the presence of tetragonal polymorph of zirconia. The same process was also utilized for preparation of YSZ powder which yields nanocrystalline cubic ZrO_2 when the precursor powder was calcined at 700°C.

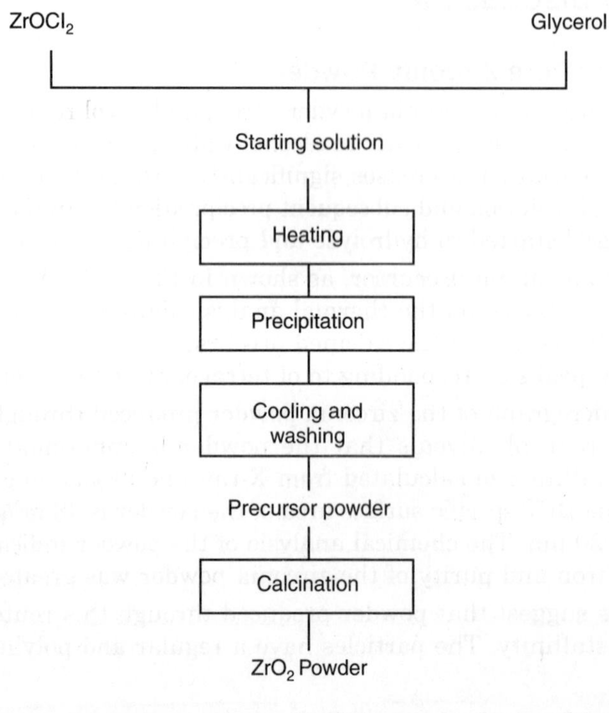

Fig. 15.1: Flow sheet of preparation of zirconia powder through glycerol route

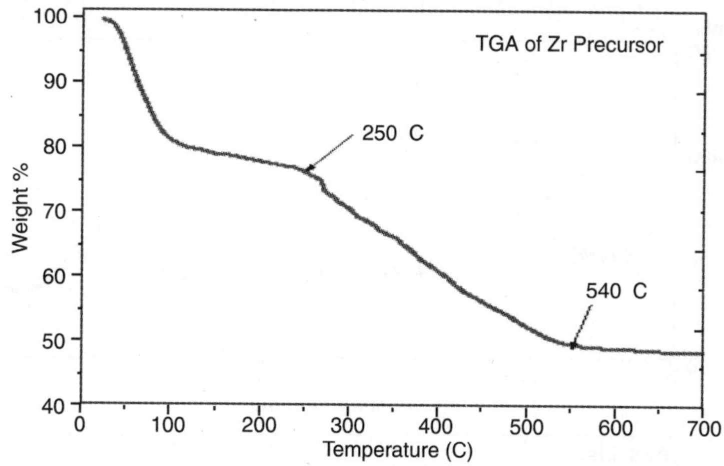

Fig. 15.2: TG plot of Zr containing precursor obtained through glycerol route

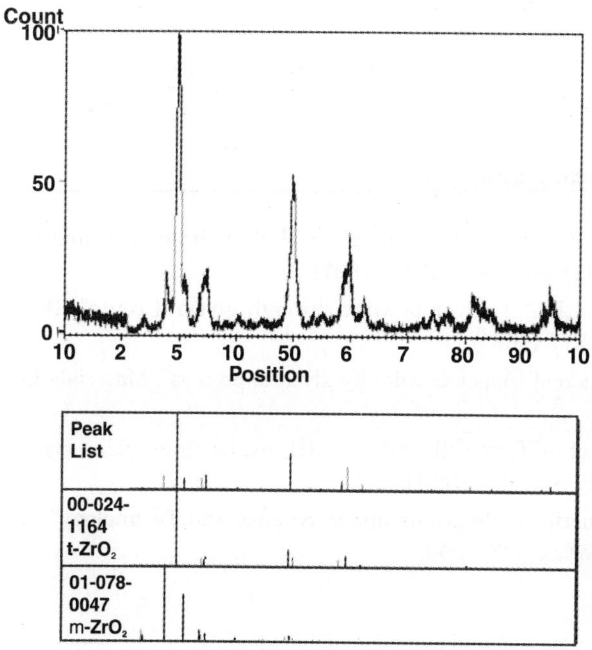

Fig. 15.3: XRD pattern of Zr-containing precursor calcined at 650°C for 2 h.

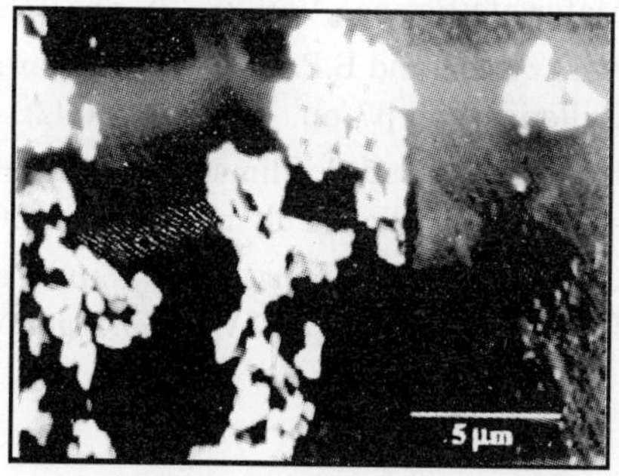

Fig. 15.4: SEM Photomicrograph of zirconia powder calcined at 650°C for 2 h

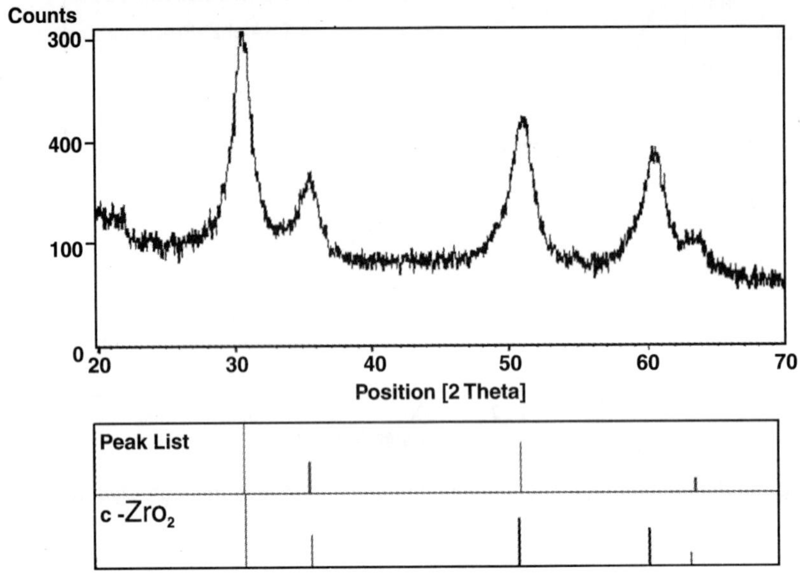

Fig 15.5: XRD pattern of YSZ powder calcined at 700 °C for 1 h

REFERENCES

1. Amit Sinha, and B.P. Sharma, "Novel Route for Preparation of High Voltage Varistor Powder", Materials Research Bulletin, Vol. 32 (11) pp. 1571–1579 (1997).
2. Amit Sinha, and B.P. Sharma, "Preparation of silver powder through glycerol route", Bulletin of Materials Science Vol. 28 (3), pp. 213–217 (2005).
3. Amit Sinha, and B.P. Sharma, "Preparation of copper powder by glycerol process", Materials Research Bulletin, Vol. 37 pp. 407–416 (2002).
4. H. Matijevis and S. Climas, "Formation of uniform colloidal iron (III) oxides in ethylene glycol-water solutions", Colloid Polym. Sci., Vol. 265, pp. 155–163 (1987).
5. O. Madelung, "Numerical Data and Functional Relationships in Science and Technology", Landolt-Börnstein New Series IV/6, Springer-Verlag, NY 1996.

❍❍

16

Synthesis and Characterization of Bi & Ta Doped Nanocrystalline PZT Ferroelectrics by Mechanical Activation

B. Praveen Kumar, H.H. Kumar, D.K. Kharat[a], M. Balasubramanian, and B.S. Murty[b]
[a]PZT Centre, Armament Research & Development Establishment and
[b]Department of Metallurgical and Materials Engineering, Indian Institute of Technology Madras, Chennai, India

Abstract

The nanocrystalline Bi and Ta doped lead zirconate titanate ceramics with the formula $Pb_{0.97}Bi_{0.03}(Zr_{0.53}Ti_{0.47})_{0.9925}O_3$ and $Pb_{0.985}(Zr_{0.53}Ti_{0.47})_{0.97}Ta_{0.03}O_3$, near the morphotrophic phase boundary, have been synthesized by mechanical activation route. Preliminary X-ray structural studies of the compounds show the existence of rhombohedral and tetragonal structure. A detailed study on dielectric constant (ε) was carried out by varying the sintering temperature. A microstructural study was also carried out using atomic force microscopy. The mechanical activation route offers a very attractive advantage of avoiding both calcination process and loss of volatile elements, with improved properties.

16.1 INTRODUCTION

Lead zirconate titanate ceramics (PZT) at composition closer to the morphotrophic phase boundary are of high significance due to their excellent piezoelectric and ferroelectric properties [1–3]. In general, PZT ceramics are processed by conventional solid-state reaction. This conventional method leads to high calcination temperature, non-stoichiometry, volatile lead loss, high sintering temperature, inevitable particle coarsening and high aggregation of powders. The presence of particle agglomerates will also end in poor microstructure and properties of PZT ceramics. Therefore it is required to process PZT ceramics at temperature as low as possible and many efforts have been made to evade this difficulty. One of them is to use ultra fine PZT powders to manufacture the PZT ceramic components. These PZT ceramics can be synthesized by wet-chemistry-based processing routes, including chemical precipitation [4], sol-gel process [5] and hydrothermal reaction [6]. Although many advantages are listed for wet-chemistry based methods, problems still persists. For instance, many of this wet-chemistry based powder precursors are to be calcined at temperature of 500–600°C, to get desired PZT phase. This requirement of calcination temperature makes these wet chemical methods similar to conventional solid state reaction. Also these chemical methods involve high purity chemicals which are expensive and moisture sensitive [7].

Mechanical activation is a promising fabrication technique for PZT ceramics, in which reaction between constituent oxides takes place during high energy ball milling. It is a solid-state powder processing route that includes repeated welding, fracturing and rewelding of powder particles [8]. The basic idea is to apply intensive milling in the solid-state synthesis of PZT and to skip the steps of calcination at elevated temperature and subsequent milling. Mechanical activation is a very useful process to prepare materials like dispersion strengthened high temperature alloys [9], nanocrystalline oxide powders [10], $YBa_2Cu_3O_7$ superconductors [11], ferrites [12] and ferroelectrics [13, 14].The advantages of mechanical activation route are preparation of Pb rich nanocrystalline ferroelectric materials at lower temperatures, low Pb loss, high dense structure and good dielectric properties [15].

In addition, dopants are often added to PZT to enhance the piezoelectric properties and to modify the microstructure. There are mainly two types of dopants–donor and acceptor dopants. Donor dopants cause Pb-vacancies by substituting a higher valence ion for Pb^{2+} or for $(Ti, Zr)^{4+}$. Examples of donor dopants are Nb^{5+}, Ta^{5+}, La^{3+}, Bi^{3+} and Sb^{5+}. Acceptor dopants cause O-vacancies by substituting a lower valence ion for Pb^{2+} or for $(Ti, Zr)^{4+}$. Examples of acceptor dopants are Sc^{3+}, K^+ or Fe^{3+}. Here we have attempted to study the effect of adding donor dopants, Bi^{3+} and Ta^{5+} to $Pb(Zr_{0.53}Ti_{0.47})O_3$ on the microstructure and properties [16]. In the present work, we report a simple approach to prepare PZT ceramics via mechanical activation by skipping the calcination step. Our results show a better understanding of size and dopant (Bi^{3+} and Ta^{5+}) effects on the piezoelectric and dielectric properties of nanocrystalline PZT ceramics.

16.2 EXPERIMENTAL WORK

PbO, ZrO_2 and TiO_2, with Bi_2O_3 and Ta_2O_5 powders of high purity were used as starting materials. The dry milling was carried out in a Fritsch Pulverisette 5 planetary ball milling system at 300 rpm with a ball to powder ration of 15:1, using 250 ml tungsten carbide vials and balls as a milling medium and for a period of 20 h. The milled powders were then uniaxially pressed into 10 mm diameter pellets at a pressure of 150 MPa. Sintering of the pellets was carried out from 1100 to 1300°C for 30 min. The sintered samples were characterized using a 3100 Nanoscope, atomic force microscope (AFM). X-ray diffraction analysis of sintered samples was performed using an AXS D8 Discover, Bruker X-ray diffractometer with Cu K_α radiation. For electrical measurements the samples were electroded using silver paste and poled in silicon oil bath at 80°C by applying a DC electric field of 3 kV/mm for 30 min. Dielectric and piezoelectric properties of Ag electroded pellets were measured using HP 4192 A impedance analyzer and take control PM 35 dual range piezometer system respectively.

Results and Discussion

Figure 16.1 shows the X-ray diffraction patterns of Bi and Ta doped (3 mol.%) PZT sintered at 1250°C for 30 min. Inclusion of Bi^{3+} into the perovskite ABO_3 structure of PZT is assumed to partially substitute at Pb-site (refer Table 16.1) and Ta^{5+} substitutes for (Zr^{4+}, Ti^{4+}) ions, by this means the charge compensation is done by cpreation of Pb^{2+} vacancies (V_{pb}). In the current work, direct precedence is given to the cationic charge vacancy compensation mechanism, as the raw materials were purposely mixed in quantities so as to satisfy the reaction. XRD studies revealed that Bi^{3+} doping at A-site and Ta^{5+} doping at B-site on morphotrophic phase boundary (MPB) of PZT resulted in ferroelectric tetragonality. In principle, at morphotropic phase boundary in PZT, both tetragonal and rhombehedral phases co-exist. The Zr-rich region in the phase diagram is rhombohedral while the Ti-rich region is tetragonal. Addition of dopants can alter the phase stability. In comparison to the literature study [18,19] our results indicate that addition of Bi and Ta resulted in intensified tetragonality (200) as can be observed from the compared literature.

Table 16.1: Ionic radius of different ions [17]

Ions	Ionic radii (nm)
Pb^{2+}	0.149
Bi^{3+}	0.116
Zr^{4+}	0.089
Ti^{4+}	0.074
Ta^{5+}	0.068

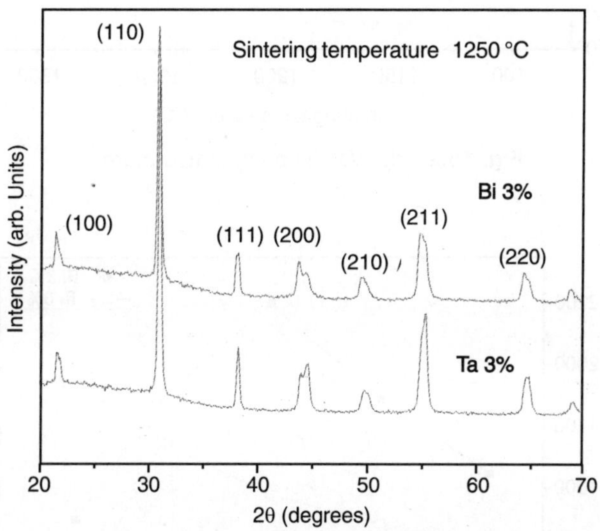

Fig. 16.1: XRD patterns of Bi and Ta (3 mol %) doped sintered PZT

Figure 16.2 depicts piezoelectric charge coefficient (d_{33}) as a function of sintering temperature for mechanically activated and sintered samples of Bi and Ta doped PZT. Piezoelectric charge coefficient increases with increase in sintering temperature up to an optimum level and then decreases. And also Ta doped PZT ceramics shows higher piezoelectric charge coefficient than Bi doped PZT. There are several parameters influencing the properties of piezoelectric materials *viz.*, homogeneity, density, grain size and different cations and its concentration etc. However, we understand that density, grain size and doping positions plays a striking effects on piezoelectric properties. According to space-charge theory [20], there are a few space-charge sites such as lattice vacancies or impurities bounding inside grain boundaries and domain walls and forming the space-charge fields which inhibit the movement of domain walls. As a ceramic becomes denser, it is expected that the space-charge fields will decrease because this field is eliminated by the space-charge of surrounding grains more easily than in less denser ceramics. When the grain size increases, the surface area of the space-charge layer will decrease, so the space-charge field will decrease simultaneously. Piezoelectric properties are also sensitive to position of dopant substitutions and formation of solid solutions. It reveals that substitution of Bi^{3+} and Ta^{5+} to A and B sites, leads to partial and complete substitution because of their size differences in ions. This ionic size difference may results in variation of distortion/stress effect which may affect the piezoelectric properties partially.

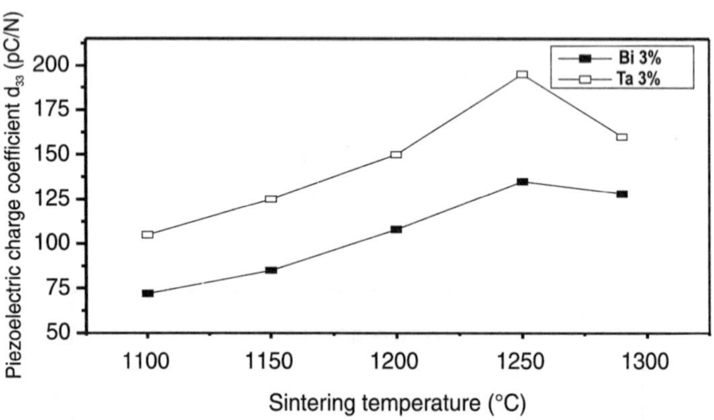

Fig. 16.2: d_{33} Vs Sintering temperature

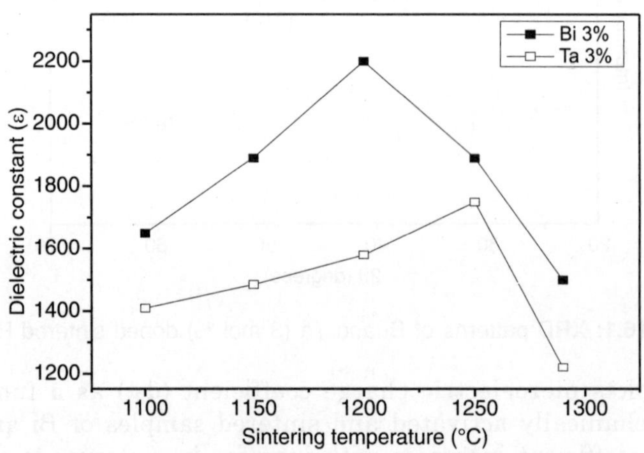

Fig. 16.3: Dielectric constant Vs Sintering temperature

Figure 16.3 shows room temperature dielectric constant of mechanically activated and sintered,

Bi and Ta doped PZT ceramics with variation of sintering temperature. It indicates that Bi^{3+} and Ta^{5+} ions doped with PZT increases the dielectric properties in relation to the pure material, because of the presence of lead vacancies. This can be attributed to the fact that the lead vacancies produced will increase the mobility of domain walls as a consequence of the drop in the stress of the lattice. A reduction of crystallite size into nanometer scale changes the domain structure and consequently the dielectric behaviour of ferroelectric materials. A local change in microstructure within nanoscale affects the type and intensity of the dipole and hence dielectric constant. The increase in dielectric constant with respect to reduced grain size is due to increase in the number of defect and electric dipoles. And also dipole-dipole interaction depends on the location of dipoles (grain core or grain boundary). While the grain size of the confined system is reduced, the fraction of dipoles at the interfaces increases significantly and dielectric constant is increased. The peak dielectric constant was observed in Bi doped PZT at 1200°C, compare to Ta doped PZT at 1250°C. This is because Bi_2O_3 in PZT forms a low melting phase that helps in liquid phase sintering, which reduces the sintering temperature and also improves the densification. This may be the reason for observation of increased dielectric constant in Bi doped samples.

Figure 16.4(a) and (b) shows the atomic force microscopy (scanning mode) images of mechanically activated and sintered Bi and Ta doped PZT samples. AFM images of the sintered PZT samples reveal the grain size of around 200 nm. These images confirm the nanocrystalline grain size, which can be one of the reasons for superior piezoelectric and dielectric properties. And it can be correlated with our above discussions.

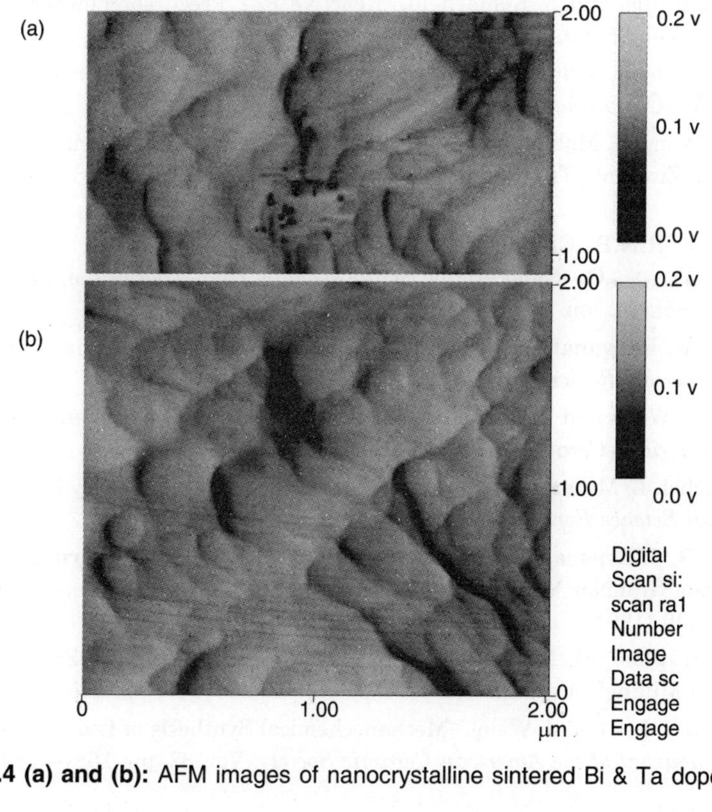

Fig. 16.4 (a) and (b): AFM images of nanocrystalline sintered Bi & Ta doped PZT

16.3 CONCLUSIONS

The nanocrystalline size, the lead vacancies, the doping sites and the charge compensation play an important role in the compositional and microstructural homogenization of Bi and Ta doped mechanically activated PZT ceramics. Piezoelectric studies conclude that Ta doped PZT shows better piezoelectric properties because of its position of dopant substitution, and its ionic size differences. Introduction of Bi in PZT lattice remarkably increases the dielectric constant and it observed in lower sintering temperature which could be attributed to the process of liquid phase sintering and increase in densification. Both piezoelectric and dielectric properties were improved with increase of densification because of its lower particle size. AFM studies reveal the presence of nanocrystalline grain size after sintering. This nanocrystalline Bi and Ta doped PZT, having superior properties in relation to the undoped material, may be suitable for sensors and actuation applications.

REFERENCES

1. B. Jaffe, W.R. Cook and H. Jaffe, *Piezoelectric Ceramics*, Academic Press Inc., New York (1971).
2. H. Heartling, "Ferroelectric Ceramics: History and Technology", *Journal of the American Ceramic Society,* Vol. 82 (4), pp. 797–818 (1997).
3. S.S. Chandreya, M. Fulath and A. Pask, "Reaction Mechanisms in the Formation of PZT Solid Solution", *Journal of the American Ceramic Society,* Vol. 64 (7), pp. 422–425 (1981).
4. K.R.M. Rao, A.V.P. Rao and S. Komarneni, "Reactive PZT Precursor Powder by Coprecipitation", *Materials Letters*, Vol. 28, pp. 463–467 (1996).
5. H. Hirashima, E. Onishi and M. Nakagowa, "Preparation of PZT Powders from Metal Alkoxides", *Journal of the Non-Crysalline Solids,* Vol. 121, pp. 404–406 (1990).
6. S.F. Wang, Y.R. Wang, T. Mahalingam, J.P. Chu and K.U. Lin, "Characterization of Hydrothermally Synthesized Lead Zirconate Titanate (PZT) ceramics", *Materials Chemistry and Physics*, Vol.87, pp. 53–58 (2004).
7. S.K.S. Parashar, R.N.P. Choudhary and B.S. Murty, "Ferroelectric Phase Transition in $Pb_{0.92}Gd_{0.08}(Zr_{0.53}Ti_{0.47})_{0.98}O_3$ Nanoceramic Synthesized by High-Energy Ball Milling", *Journal of the Applied Physics*, Vol. 94, pp. 6091–6096 (2003).
8. B.S. Murty and S. Ranganathan, "Novel Materials Synthesis by Mechanical Alloying/Milling", *International Materials Review,* Vol. 43, pp. 101–144 (1998)
9. N.J. Welham, P.E. Willis and T. Kerr, "Mechanochemical Formation of Metal-Ceramic Composites", *Journal of the American Ceramic Society,* Vol. 83, pp. 33–40 (2000).
10. E. Gaffet, D. Michel, L. Mazerolles and P. Berthet, "Effects of High energy Ball Milling on Ceramic Oxides", *Material Science Forum*, Vol. 235–238, pp. 103–108 (1997).
11. M. Simoneau, G. L'Esperance, M.L. Trudeau and R. Schulz, "Structural and Magnetic Characterization of Granular $Y_1Ba_2Cu_3O_{7-d}$ Nanocrystalline Powders", *Journal of Materials Research*, Vol. 9, pp. 535–540 (1994).
12. G. Nicoara, D. Fratiloiu, M. Nogues, J.L. Dormann and F. Vasiliu, "Ni-Zn Ferrite Nanoparticles Prepared by Ball Milling", *Material Science Forum,* Vol. 235–238, pp. 145–150 (1997).
13. J. Xue, D. Wan, S.E. Lee and J. Wang, "Mechanochemical Synthesis of Lead Zirconate Titanate from Mixed Oxides", *Journal of the American Ceramic Society,* Vol. 82, pp. 1687– 1692 (1999).

14. B. Praveen Kumar, H.H. Kumar, D.K. Kharat and B.S. Murty, "Investigation and Characterization of La-doped PZT Nanocrystalline Ceramic Prepared by Mechanical Activation Route", *Materials Chemistry and Physics*, Vol. 112, pp. 31–34 (2008).
15. L.B. Kong, J. Ma, W. Zhu and O.K. Tan, "Highly Enhanced Sinterability of Commercial PZT Powders by High-Energy Ball Milling", *Materials Letters,* Vol. 46, pp. 274–280 (2000).
16. Md. A. Mohiddon, A. Kumar, K.L. Yadav, Effect of Nd Doping on Structural, Dielectric and Thermodynamic Properties of PZT (65/35) Ceramic, *Physica B,* Vol. 395, pp. 1–9, (2007).
17. K. Ramam, and L. Salinas, "Effect of Ba on Ferroelectric and Piezoelectric Properties of the PLZT (1.2/55/45) System", *Physical Status Solidi (a)* Vol. 203, pp. 2119–2127 (2006).
18. K.B. Lee and H.S. Lee, "Characterization of a Bismuth-Doped Lead Zirconate Titanate Thin Film Capacitor", *Journal of the Korean Physicsl Society*, Vol. 31, pp. 532–536, 1997.
19. A. Banerjee, A. Bandyopadhyay and S. Bose, "Influence of La_2O_3, SrO and ZnO Addition on PZT", *Journal of the American Ceramic Society,* Vol. 89, pp. 1594–1600 (2006).
20. C.A. Randall, N. Kim, J.P. Kucera, W. Cao and T.R. Shrout, "Intrinsic and Extrinsic Size Effects in Fine-Grained Morphotropic-Phase-Boundary Lead Zirconate Titanate Ceramics", *Journal of the American Ceramic Society,* Vol. 81, pp. 677–688 (1998).

❍❍

17

Synthesis and Characterization of $CaWO_4$: Ag Powders for Electronic Display Applications

U. Rambabu, N.R. Munirathnam, and T.L. Prakash
Centre for Materials for Electronics Technology (C-MET), Hyderabad, India

Abstract

This paper reports on the synthesis of $Ca_{(1-x)}WO_4 : Ag_x$ (0 d" x d" 0.05 mol %) powder phosphors by a hydrothermal method and followed by calcinations. We have investigated the effects of Ag-content on photoluminescence (PL) spectra of $CaWO_4$. The phase nature of the synthesized powder phosphors have been examined from X-ray diffractometry. Thermal decomposition behaviour of the dried $CaWO_4$ powder phosphor have been studied from TG/DTA profiles and thus noticed that there exists a polymorphic transition at 800 °C. Based on the PL results, the dopant Ag content in the phosphors has been optimized to be 0.001 mol %, displaying a prominent *blue* emission from the phosphor. From the FT-IR spectrum, a strong peak at 845 cm^{-1} has been obtained due to the stretching vibration of WO_4^{2-} in scheelite structure, and a weak but sharp band at 445 cm^{-1} has also been noticed due to the metal, oxygen (Ca-O) band. Scanning Electron Microscopic (SEM) studies reveal that the synthesized $CaWO_4$ and Ag-doped $CaWO_4$ particles are in plate shaped in the range from 400 nm to 1 micron. From SEM images, it has been noted that the agglomeration of synthesized powder phosphors has been increased with the change in Ag content, owing to the density variation. The particle size distribution of the synthesized powder phosphors has also been studied.

17.1 INTRODUCTION

Multi-metal oxide ceramics are found to be more promising materials for different optical and electronic applications [1-2]. Such oxides based phosphors have widely been used in the development of CRTs (Projection TVs), field emission vacuum fluorescent, electroluminescent displays and scintillators in X-ray and positron emission tomographs [3]. The specific luminescent properties of metal oxide ceramics are highly influenced by the host material, dopant ion concentration, chosen stoichiometry and also the preparation conditions [4]. Rare earth ions doped tungstate crystals prepared by different methods have received a great deal of attention as promising laser materials, due to their attractive third order nonlinear dielectric susceptibilities. Tungstates are divided into two groups as Scheelites ($CaWO_4$, $BaWO_4$, $SrWO_4$ and $PbWO_4$) and as Wolframites ($MgWO_4$, $ZnWO_4$ and others). As a representative example, $CaWO_4$ with the scheelite structure, containing Ca^{2+} ions and WO_4^{2-} groups with the coordination no of eight for Ca^{2+} and four for W^{6+} is considered to be highly functional material due to its intriguing luminescence properties. It mainly shows an efficient blue emission when excited by UV, X-rays and cathode rays. Owing to its wide variety of applications, several new routes of synthesis of pure and homogenously doped calcium tungstates are found to be significantly relevant [5–12]. Zhang et el. have reported that, ZnS: Ag at high current densities could show brightness saturation due to the ground state depletion or to non radiative Auger process at low Ag content. It has also been reported that activator recycling may not be possible due to its relatively high decay time ($\tau_{1/e} \sim 25$ μs) [13]. Kransnov al et. have noticed that the presence of Ag as a co-dopant has improved the luminescence efficiency in the case of ZnS: Mn phosphors by reducing the Mn clustering effects [14]. Mostly $CaWO_4$ synthesis has been carried out in the form of single crystal powders and films [8], by precipitation method [15], traditional solid state reaction [16], and spray pyrolysis route [17-18]. In the present paper, we report on yet another method of synthesis, characterization and the influence of dopant Ag content on the blue emission performance of $Ca_{(1-x)}WO_4$: Ag_x ($0 \le x \le 0.05$ mol %) powder phosphors.

17.2 EXPERIMENTAL

$CaWO_4$: Ag powder phosphors were prepared by a hydrothermal method with the following possible chemical reaction:

$CaCl_{2\,(1-x)} + Na_2WO_4 + x\,AgNO_3 \rightarrow CaWO_4\text{: Ag} + 2\,NaCl + NO_2\uparrow$ (Where x = 0, 0.001, 0.005, 0.01 and 0.05 mol %) $\rightarrow$ {1}

The precursor solutions of the $CaCl_2$, Na_2WO_4 and $AgNO_3$ were prepared by dissolving (in mol %) the respective G.R grade chemicals. Sodium tungstate solution was added to the $CaCl_2$ solution under a vigorous stirring. Ethylene glycole was used as a dispersing media. To prepare Ag-doped $CaWO_4$ powder phosphors, $AgNO_3$ solution was added by a drop wise to the above $CaWO_4$ resultant dense media with continuous stirring by magnetic stirrer. The P^H of the above resultant dense liquid was maintained at 6.5–7.0 by using HNO_3 and NaOH with stirring for 3 hrs. The resultant precipitated mass was transferred into a 100 ml capacity teflon lined SS made auto-clave, which was heated in a electrical oven at 175 °C for 24 hrs under a self generated pressure. After a 24 hrs heating of the autoclave vessel without any stirring or disturbance, it was gradually cooled down to the room temperature. The resultant precipitated mass was separated centrifugally and washed thoroughly with DI water to remove the traces of NaCl and ethylene glycole. It was later dried at 120 °C for few minutes in an electrical oven. Thus the obtained fluffy powder was taken into a quartz boat and calcined in a resistance heated tubular furnace starting from room temperature to 800 °C and was kept at this temperature for 2 hrs under ambient atmosphere. The body colour of the synthesized powder phosphors were observed to be changing from white to magenta colour with an increased silver content. The synthesized powder phosphors were displaying blue emission under an UV source.

The crystalline and phase purity of the synthesized $CaWO_4$: Ag powder phosphors were studied on a Phillips, X-Ray Diffractometer. XRD pattern of each phosphor sample was recorded in the range of diffraction angle $2\theta = 5 - 80^o$. Thermal decomposition behaviour of $CaWO_4$ powder phosphor was studied by using Mettler Toledo Star SW 9.01, TG/DTA. The dried 3.4688 mg of $CaWO_4$ powder was heated from room temperature to 1100 °C @10 °C per minute under argon atmosphere. FT-IR spectrum of the synthesized powder phosphor was measured on a Perkin Elmer, FT-IR spectrometer by KBr pellet technique. Particle shape, size and surface morphology of the obtained powder phosphors were studied on a Phillips XL 30, scanning electron microscope (SEM). The particle size distribution of the Ag-doped and undoped powder phosphors were also studied by using Nicomp 380, submicron particle sizer. The sample solutions were prepared by using diethyl ether solvent and a Sonics–Vibra cell Sonicator. Photoluminescence spectra of the synthesized powder phosphors were measured by using an excitation wavelength, $\lambda_{exc} = 261$ nm, on a Hitachi, 650-10S Spectrofluorimeter.

17.3 RESULTS AND DISCUSSION

Based on d-spacing (Å), relative intensity and diffraction angle (2θ) of XRD profile (Fig.17.1), it is conformed that the obtained powders were in scheelite structure *i.e.*, CPDF:41-1413

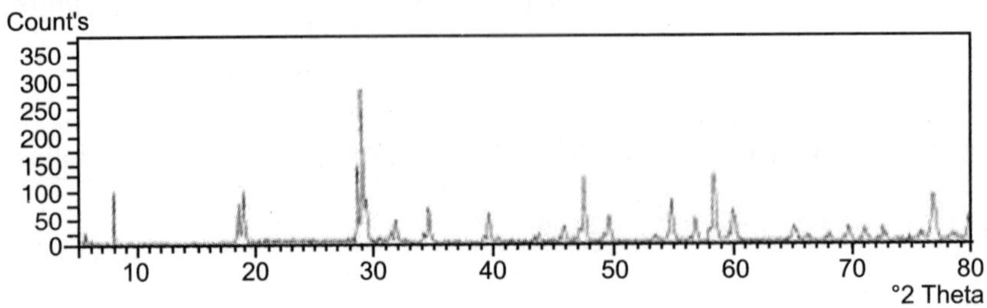

Fig. 17.1: XRD profile of $CaWO_4$ powder phosphor

TG/DTA curves [Fig. 17.2] show that the decomposition starts below 300°C with a loss of water from the coordination sphere of the sample. The thermal decomposition behaviour has been attributed to the endothermic and exothermic effects in the DTA curve. A weak exothermic peak at 340°C indicates the total decomposition of the traces of the solvent ethylene glycole. It is clear to realize that TG curve shows a 10% of weight loss in the sample while heating from room the temperature to 500°C under an Ar gas atmosphere. From 500–900°C there has not been any change in the weight of the sample. A weak exothermic peak at 800°C could possibly because of a polymorphic transition of $CaWO_4$. Thus, the thermal characterization of $CaWO_4$ provides information on the mechanism involved in the thermal decomposition of the Ca–W–O precursor and the optimum conditions for the final annealing can also be determined. According to the TG/DTA analysis data, the final annealing temperature for the preparation of $CaWO_4$ could be around 800°C [19].

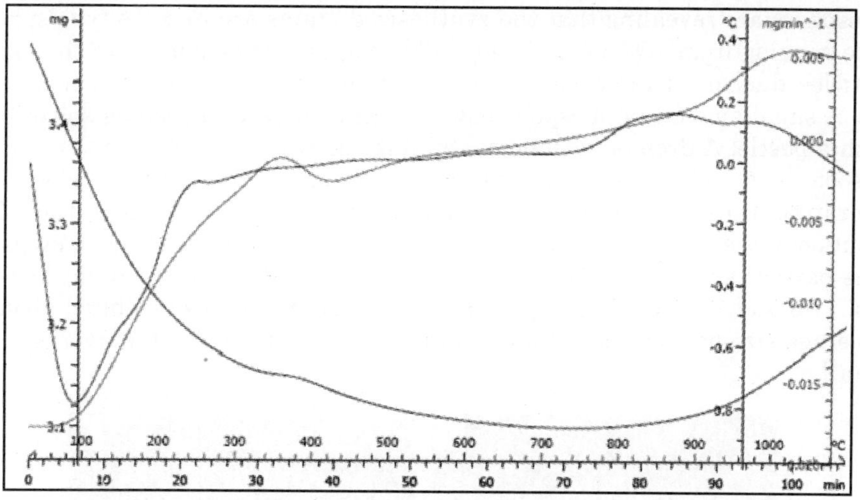

Fig. 17.2: TTG/DTA/TG profiles of $CaWO_4$ powder phosphor

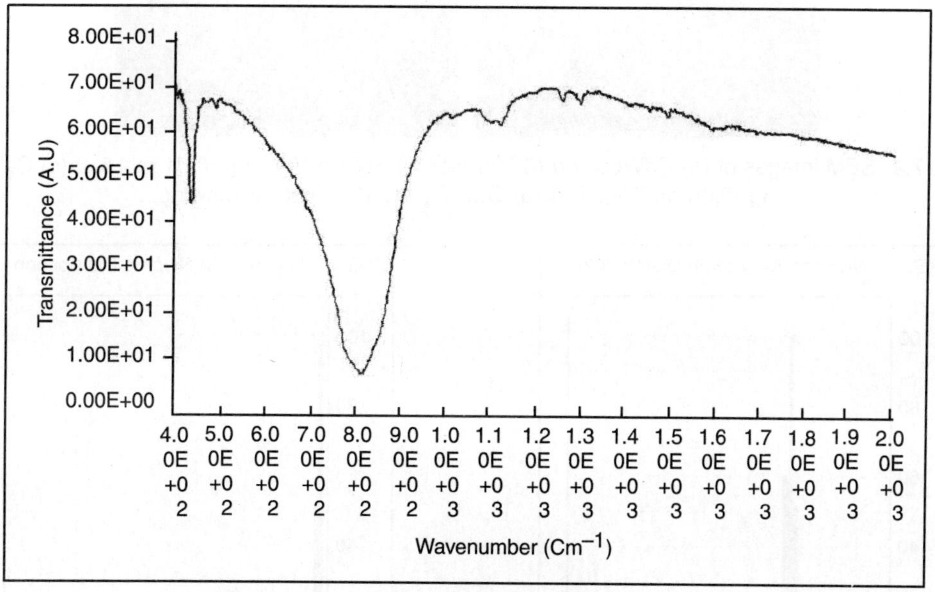

Fig. 17.3: FT-IR spectrum of $CaWO_4$ powder phosphor

FT-IR spectrum shows a strong peak at 845 cm^{-1} which is due to the characteristic stretching vibration of WO_4^{2-} in scheelite. A sharp peak at 445 cm^{-1} indicates the bending vibration of metal-oxygen (Ca–O) bond. Two weak peaks at 1170 and 1370 Cm^{-1} are due to the H-OH stretching vibration [20]. Scanning electron micrographs (Fig. 17.4) of Ag: $CaWO_4$ and $CaWO_4$ powder phosphors are revealing that the synthesized grains are in plate type structures with particle size ranging from 400 nm to 1 μm. SEM images also demonstrate the agglomeration of the particles due to an increase in Ag content in the phosphor studied due to density differences. A small amount of phosphor powder was dispersed in acetone solvent in an agate mortar with a pestle. A drop of acetone containing the sample was placed on a SEM sample holder, as soon as the solvent acetone evaporates most of the phosphor particles were found to be agglomerating, thus an increase of Ag-content increases the agglomeration. The particle size distribution of the synthesized powder phosphors were also studied by using particle sizing system. The particle size distribution with reference to the doped Ag- content has been shown in the Figs. 17.5 and 17.6. $CaWO_4$ doped with 0.005 mol % Ag has exhibited about 97.96% of 42.6 nm average size particles and the remaining 2.04% with 596.3 nm average particle size [Fig. 17.5 (a)].

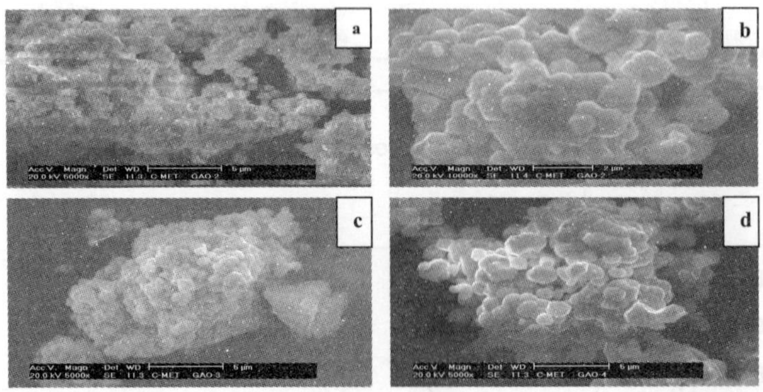

Fig. 17.4: SEM images of (a) $CaWO_4$: Ag (0.005 mol %), (b) $CaWO_4$: Ag (0.01 mol %), (c) $CaWO_4$: Ag (0.05 mol %), and (d) $CaWO_4$: Ag (0.0 mol %) powder.

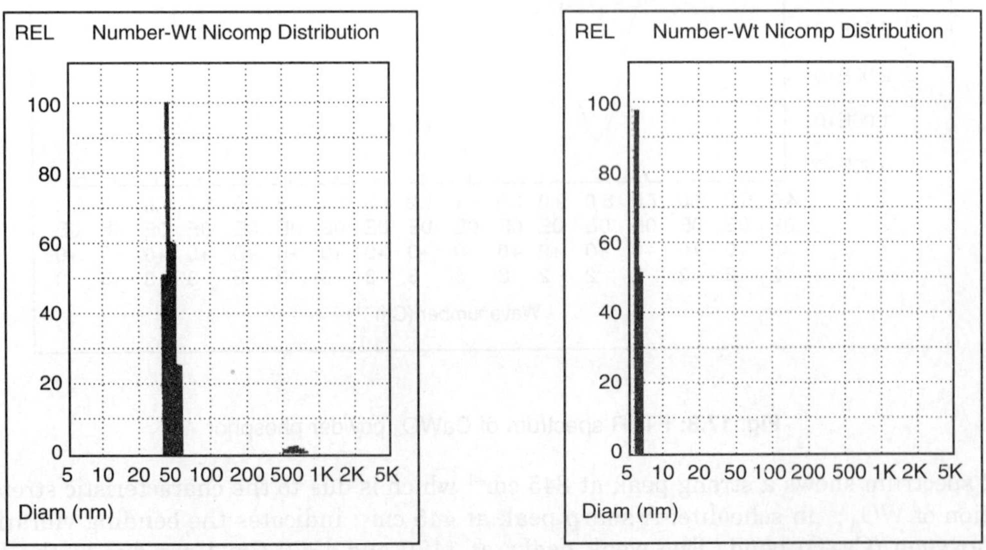

(a) (b)

Fig. 17.5: Particle size distribution of (a) $CaWO_4$: Ag (0.005 %) and (b) $CaWO_4$: Ag (0.01 %) powder phosphors.

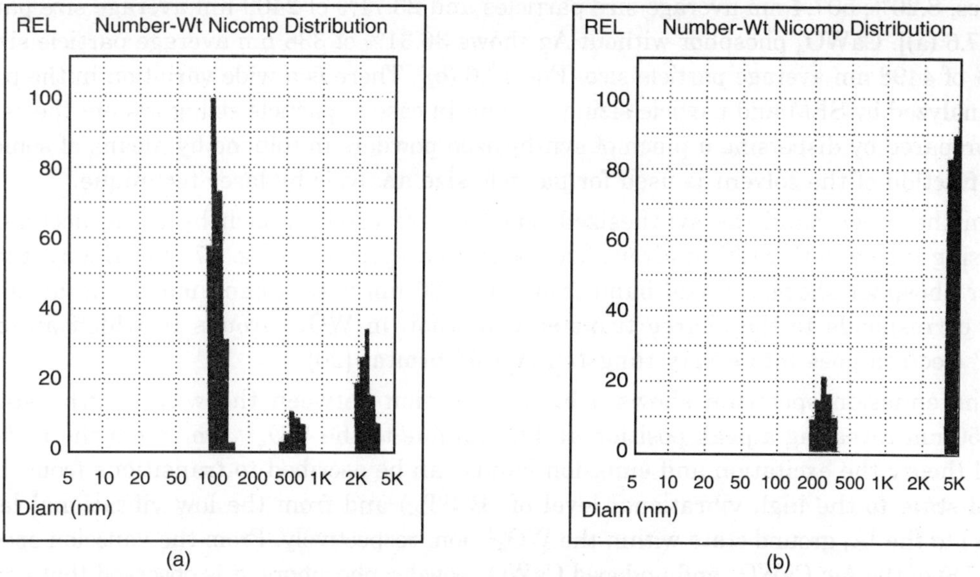

Fig. 17.6: Particle size distribution of (a) CaWO$_4$: Ag (0.05 %) and (b) CaWO$_4$: Ag (0.0 %) powder phosphors.

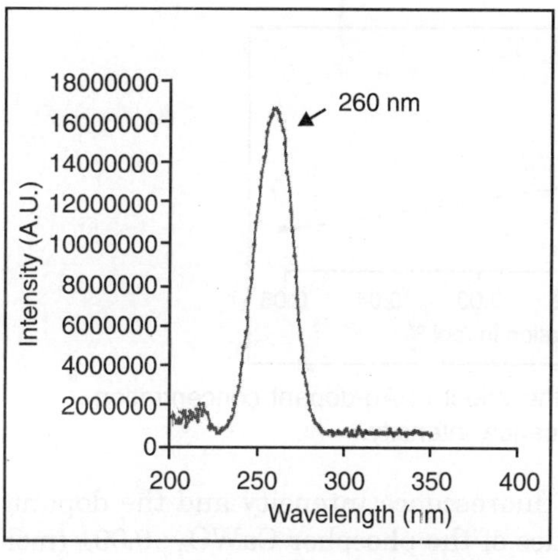

Fig. 17.7: Excitation spectrum of CaWO4: Ag (0.001 %) powder phosphor.

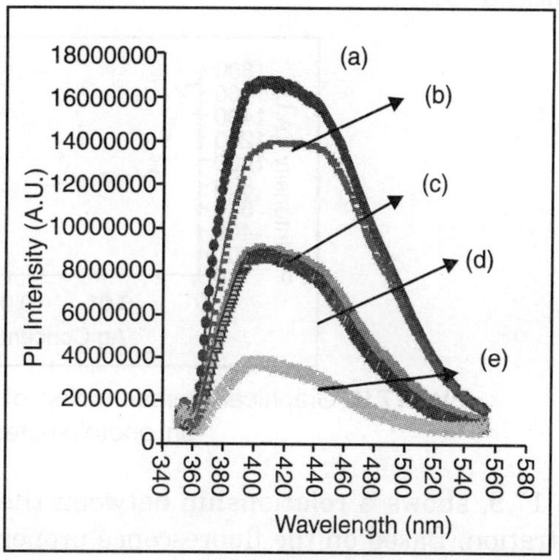

Fig. 17.8: Photoluminiscence Spectrum of
(a) CaWO$_4$ Ag 0.01mol%
(b) CaWO$_4$ Ag 0.05 mol%
(c) CaWO$_4$ Ag 0.0011mol%
Powder Phosphors

It has been noticed that the CaWO$_4$ containing 0.01 mol% of Ag is having 100% 5–9 nm average particles [Fig. 17.5 (b)]. CaWO$_4$ with 0.05 mol% of Ag is having 68.09%; 97.3 nm average size particles, 8.20%; 507.4 nm average size particles and 23.71% of 2409 nm average size particles [Fig. 17.6 (a)]. $CaWO_4$ phosphor without Ag shows 36.31% of 336 nm average particle size and 63.69% of 4498 nm average particle size [Fig. 17.6 (b)]. There is a wide variation in the particle sizes analyzed by SEM and particle sizing system. In case of particle sizing system the samples were prepared by dispersing a pinch of synthesized powders in toluene by means of sonication and a fraction of the solvent is used for particle size analysis by laser technique.

On the other hand the synthesized powders are used as such by using acetone as a dispersing media with SEM analysis. The excitation spectrum (Fig. 17.7) of the synthesized powder phosphor shows a broad band from 240–280 nm with a band maximum at 261 nm, which corresponds to the charge transfer transition in WO_4^- groups in which an oxygen (O) 2P electron goes into empty tungsten (W) 5d orbital [20].

The emission spectrum shows a broad emission between the wavelength ranges of 360–550 nm revealing a peak position at 420 nm due to the WO_4^{2-} ion. From the molecular orbital theory the excitation and emission bands can be ascribed to transitions from the $^1A_1$ ground state to the high vibrational level of ^1B ($^1T_{2-}$) and from the low vibrational level of ^1B ($^1T_{2-}$) to the $^1A_1$ ground state within the WO_4^{2-} ion, respectively. From the emission spectrum (Fig. 17.8) of the Ag: $CaWO_4$ and undoped $CaWO_4$ powder phosphors, it is observed that emission band shifts towards the lower wavelength for the Ag: $CaWO_4$ compared with the undoped $CaWO_4$ phosphor. Silver at lower concentration levels has been showing a sensitized luminescence. Dopant ion (Ag) concentration quenching effect in $CaWO_4$ scheelite type powder phosphor has been noticed [21].

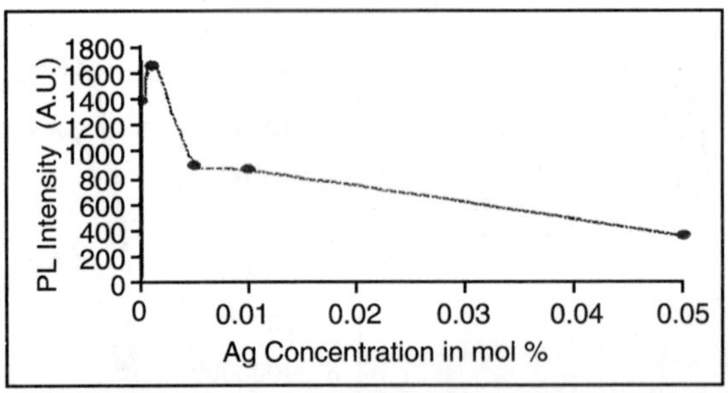

Fig. 17.9: Graphical representation of the effect of Ag-dopant concentration on photoluminescence intensity.

Figure 17.9, shows a relationship between the fluorescence intensity and the dopant Ag ion concentration. Based on the fluorescence properties of the phosphor $CaWO_4$: 0.001 (mol %) Ag, which could be suggested as a novel optical material of technological importance in the progress of phosphor coated screens of certain electronic systems.

17.4 CONCLUSIONS

By employing a hydrothermal method, a strongly *blue* luminescent Ag-doped $CaWO_4$ powder phosphor has successfully been synthesized and carried out a systematic analysis based on the features obtained from the XRD, TG/DTA, FT-IR and photoluminescence spectra. Thus the results obtained in the present study clearly demonstrate the luminescent potentiality of the optical material. Since the results are encouraging to suggest that $CaWO_4$: Ag (0.001mol %), as a promising *blue* light emitting phosphor optical system for its application in certain electronic displays.

Acknowledgements

The authors would like to thank Dr. P. K. Khanna, C-MET, Pune for extending his cooperation towards particle size distribution analysis.

REFERENCES

1. R. Manalert and M. N. Rahaman, J. Mater. Sci. 31 (1996) 3453.
2. E. Comini, A. Toncelli, M. Tonelli, E. Zannoni, E. Cavalli, A. Speghini and M. Beltinelli, J. Opt. Soc., Am. B, 14 (1997) 1938.
3. N.V. Kuleshov, A.A. Lagatsky, A.V. Podlipensky, V.P. Mikhailov and G. Huber, Opt. Lett. 22 (1997)1317.
4. A. Katelnikovas, L. Grigorjeva, D. Millers, V. Pankratov and A. Kareiva, Luthuanian, J. of Phys. 47(1) (2007) 63.
5. V. N. Kolobanov, I.A. Kamenskikh, V.V. Mikhailin, I. N. Shpinkov, D.A. Spassky, B.I. Zadneprovsky, L. I. Potkin and G. Zimmerer, Nucl. Instrum. Methods A 486 (2002) 496.
6. V. Nagirnyi, E. Feldbach, L. Jonsson, M. Kirm, A. Lushchik, C. Lushchik, L.L. Nagornaya, V.D. Ryzhikov, F. Savikhiu, G. Svensson and I.A. Tupitsina, Radiat. Meas. 29 (1998) 247.
7. P. Mogilevsky, T.A. Parthsarathy and M.D. Petry, Acta Materialia, 52 (2004) 5529.
8. M. Nikl, P. Bohacek and E. Mihokova, J. Appl. Phys. 91 (2002) 5041.
9. M. Nikl, P. Bohacek, E. Mihokova, J. Lumin. 87–89 (2000) 1136.
10. M. Kobayashi, M. Ishii, Y. Usuki and H. Yahagi, Nucl. Instrum. Meth. Phys. Res. A 333 (1993) 429.
11. G. Zhang, R. Jia and Q. Wu, Mater. Sci. Engg. B 128 (2006) 254.
12. J.H. Ryu, S. Y. Bang, W. S. Kim, G.S. Park, K.M. Kim, J.W. Yoon, K.B. Shim, and N. Koshizaki, J. Alloys Comps. 441 (2007) 146.
13. F.L. Zhang, S. Yang, C. Stoffers, J. Penczek, P.N. Yocom, D. Zaremba, B.K. Wagner and C.J. Summers, Appl. Phys. Lett. 72 2226 (18).
14. A.N. Krasnov, J.P. Bender and W.Y. Kim, Thin Sol. Films 467 (2004) 247.
15. E.F. Paski, M.W. Blades, Anal. Chem. 60 (11) (1988) 1224.
16. G. Blasse, L.H. Brixner, Chem. Phys. Lett. 173 (1990) 409.
17. Z.D. Lou, M. Cocivera, Mater. Res. Bull., 37 (2002) 1573.
18. S.J. Chen, J. Li, X.T. Chen, J.M. Hong, Z. Xue and X.Z. You, J. Cryst. Grow. 253 (2003) 361.
19. A. Katelnikovas, L. Grigorjeva, D. Millers, V. Pankratov and A. Kareiva, Lithuanian J. Phys. 47(1) (2007) 63.
20. P.Y. Jia, X.M. Liu, G.Z. Li, M. Yu, J. Fang and J. Lin, Nanotechnology 17 (2006) 734.
21. A.N. Kransnov, J.P. Bender and W.Y. Kim, Thin Sol. Films 467 (2004) 247.

❍❍

18

The Effect of Chemistry and Microstructure on Mechanical Properties of Tungsten Heavy Alloys

Jiten Das, K. Jagan Mohan, A. Chakraborty, N. Eswara Prasad, and Bijoy Sarma*
Defence Metallurgical Research Laboratory, Kanchanbagh, Hyderabad, India.

Abstract

Several tungsten heavy alloys based on W-Ni-Cu and W-Ni-Fe were made recently at DMRL by liquid phase sintering route. These alloys were heat treated and characterized in terms of density, hardness and tensile properties. Although the densities of W-Ni-Cu based heavy alloys were found to be relatively higher than those of W-Ni-Fe based heavy alloys, the former exhibited inferior tensile properties and hardness values than the latter alloys. In both these heavy alloys, un-melted solid tungsten particles were bonded together by the respective low melting matrix alloy. The strength of the matrix or the bonding strength of the interface (between tungsten particles and matrix) was found to control the nature of final fracture. Poorer matrix strength or interfacial strength was found to initiate the fracture by separation of tungsten particles either by matrix failure or by interface failure. On the other hand, tensile fracture takes place predominantly by cleavage fracture of tungsten particles, if both the matrix and interface are stronger than the tungsten particles. The tensile fractured surfaces were examined under scanning electron microscope (SEM). Failure stress is observed to be lower in case of separation of tungsten particles (due to matrix and interface failure) than that in case of cleavage fracture (transgranular fracture) of tungsten particles. SEM image of fractured surfaces clearly indicated that failure in case of W-Ni-Cu based heavy alloys was due to matrix and/or interface (between tungsten grain and matrix) failure, where as, W-Ni-Fe based heavy alloys were failed predominantly by cleavage fracture of tungsten particles. This was evidenced by poorer tensile properties obtained for W-Ni-Cu based heavy alloys as compared to those of W-Ni-Fe based heavy alloys. The present set of results indicate that heavy alloys based on W-Ni-Fe and

W-Ni-Cu can be strengthened by minor additions of Co, Mo and Fe selectively. Higher strength (> 850 MPa) of the heavy alloys are associated with the finer tungsten grain size and cleavage fracture of tungsten particles. Where as, the lower strength is associated with the larger tungsten grain size and interface and/ or matrix failure. It is found in this study that as the tungsten grain size decreases, the tensile fracture mode changes gradually from interface failure to matrix failure to tungsten grain cleavage failure.

18.1 INTRODUCTION

Tungsten based heavy alloys have high density, good mechanical properties and are easily machinable. Because of their good machinability, highly complicated shaped components can be made out of these alloys. As they posses very high density, the alloys can be used in a variety of applications such as kinetic energy penetrators, radiation shielding and counter weight materials [1]. Accoring to Islam *et al.* [2] volume fraction of tungsten particle, particle size of tungsten, closeness of tungsten particles affect the mechanical properties of tungsten heavy alloys. He also explained that raw materials purity also affects the mechanical properties of these alloys. However, the tensile properties, density and hardness of these alloys largely depend on the chemistry. W-Ni-Fe based heavy alloys possess far superior tensile properties and hardness as compared to the W-Ni-Cu based heavy alloys [3]. Both the Ni/Fe or Ni/Cu ratio and the amount of W content also significantly influence the tensile properties of these alloys [3]. Song *et al.* reported that heavy alloys with Ni/Fe ratio of 4:1 shows best tensile properties and impact strength [4]. Animesh Bose *et al.* [5] showed that lowering the W content of the alloys up to 90 wt.% causes improvement in ultimate tensile strength as well as elongation value of the sintered tungsten heavy alloys. They also observed that 92.5%W-(Ni-Fe-Co) alloy deformed with 95% showed a tensile strength of 1720 with concomitant elongation of 16%. Boris Katavic [6] explained that the 'mechanical properties of W-Ni-Co' with 'nickel to cobalt ratios ranging from 2 to 9' are far superior to that of W-Ni-Fe alloys. Animesh Bose *et al.* [5] also showed that heavy alloys 'doped with Mo and Re' exhibit improved yield strength as compared to unalloyed heavy alloys. However, according some study, pointed out by Liu *et al.* [7], higher Mo addition to tungsten heavy alloys leads to brittleness due to formation of precipitated phase. Although Re is very effective strengthener for tungsten heavy alloys according to Bose *et al.* [5], its use is very restricted because of its cost is very high.

Alloying elements also were found to influence the failure mode of tungsten heavy alloy in tension by way of modifying matrix strength, interfacial (between tungsten grain and matrix) strength and tungsten grain strength. Liu *et al.* [7] again pointed out that interfacial strength between tungsten particle and matrix affect the toughness of these alloys significantly. The present work is thus directed to understand the different types of tensile failure mechanisms in tungsten heavy alloys and also aims to establish a correlation between the strength and failure mode in tensile loading.

18.2 EXPERIMENTAL

Four W-Ni-Cu base heavy alloys with different W content and varied Ni/Cu ratio and three W-Ni-Fe base heavy alloys with different W content as well as varied Ni/Fe ratio were prepared by liquid phase sintering at 1500°C for 1.5 h in H_2 atmosphere in a continuous pusher type furnace (FHD Furnace Limited, England). The alloys were prepared using finely divided W, Mo, Co Ni, Cu, and Fe elemental powders. Particle size distribution of W and Mo are represented in Figure 18.1.

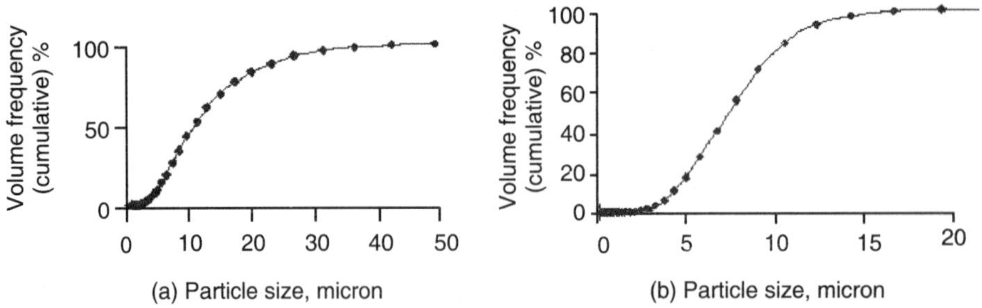

Fig. 18.1: Particle size distribution of (a) W powder [9] and (b) molybdenum powder

It is observed that molybdenum powders are finer than the tungsten powder. SEM image of tungsten powder is shown in Fig.18.2. Particle size distribution of other powders was not evaluated, because during liquid phase sintering the other elements would be completely melted and form the matrix, which binds the unmelted tungsten particles together. In the W-Ni-Fe base alloys, small amount of strengthening elements, such as Co and Mo, are added in order to vary the tensile strength of the alloys. Similarly in W-Ni-Cu base heavy alloy, little amounts of Fe is added again to obtain different tensile strength values. Our objective is to vary the strength level of these heavy alloys, so that one can establish a correlation between the mode of fracture (during tensile testing) and the obtained tensile strength values.

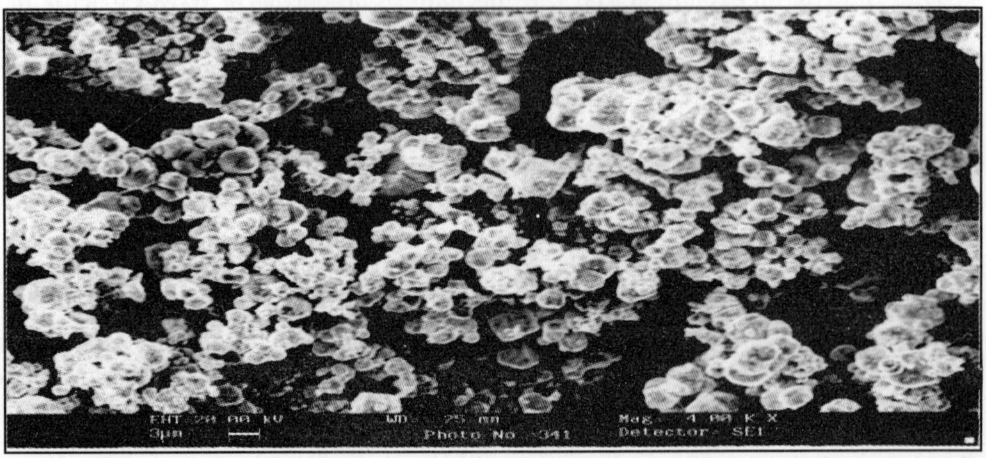

Fig. 18.2: SEM image of as-received tungsten powder utilized in this study [9]

As-sintered alloys were heat treated at 1100°C for 5 h in Argon atmosphere and subsequently characterized by evaluating their microstructure, density, hardness and tensile properties. Density of the alloys were measured by Archimedes water immersion technique. Hardness of the alloys were measured using Wolpert Vicker Hardness Tester using 30 kg load. The tensile specimens were prepared as per ASTM standard E 8 (2003) and were tested using a Masanto Tensometer. The fractured surfaces of the tensile tested specimens were observed using LEO Scanning electron microscope. EDS spectrographs from the matrix for all the alloys were also taken.

18.3 RESULTS AND DISCUSSION

Microstructure: Optical microstructure of W-90.5, Ni-7.2, Fe-1.8, Co-0.45, Mo 0.05 alloy is shown in Fig. 18.3. Around 11–15 volume% matrix and about 85–89 volume% tungsten particles are present in the microstructure. Average tungsten particle size is 30 mm.

Density, Hardness and Tensile Properties: The density, hardness and tensile properties of various W-Ni-Cu base and W-Ni-Fe base heavy alloys are given in Table 18.1. These data are the average of minimum three samples. In W-Ni-Cu base heavy alloys improvement in hardness and tensile properties were observed by minor additions (0.1wt.%) of iron. In these alloys, with a fixed Ni/Cu ratio, lower the W content, better was the observed hardness and tensile properties. However, deterioration of hardness and tensile properties with higher tungsten content could be offset to a certain extent by increasing Ni/Cu ratio. In case of W-Ni-Fe alloys, by lowering the W content up to 90.5 wt% and increasing the Ni/Fe ratio up to 4/1, improvement in tensile strength as well as elongation value was achieved.

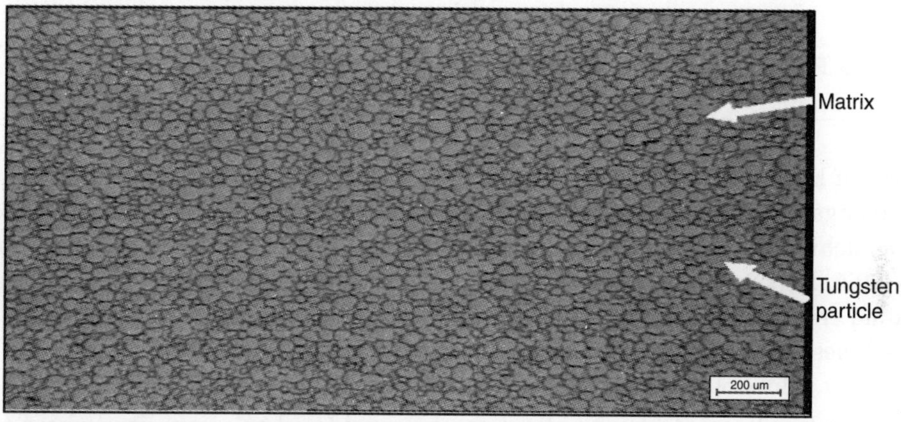

Fig. 18.3: Optical microstructure of W-90.5, Ni-7.2, Fe-1.8,Co-0.45, Mo-0.05 alloy. Vol% of matrix is 11–15, vol% of tungsten particles is 85–89% and average tungsten particle size is 30 mm

Table 18.1: Density, Hardness and Tensile properties of various tungsten base heavy alloys after sintering and heat treatment. The alloys were sintered at 1500°C for 1.5 h in H_2 atmosphere and heat treated at 1100°C for 5 h in argon atmosphere and then water quenched

Tungsten Heavy Alloy Composition (wt.%)	Density g/cc	Pore Free Density g/cc	Tensile Strength MPa	Elongation %	Hardness VHN/30kg
W-94.9, Ni-3.4, Cu 1.7 (Ni:Cu = 2:1)	17.9	18.2	655	4	276 ± 19
W-92.5, Ni-5, Cu-2.5 (Ni:Cu = 2:1)	17.6	17.7	681	6	295 ± 25
W-96.1, Ni-2.8, Cu-1.1 (Ni:Cu = 2.5:1)	18.2	18.5	670	3	286 ± 20
W-96.0, Ni-2.8, Cu-1.1, Fe-0.1 (Ni:Cu = 2.5:1)	18.1	18.5	701	5	298 ± 23
W-93.5, Ni-4.0, Fe-2.0, Co-0.5 (Ni:Fe = 2:1)	17.7	17.8	853	14	310 ± 21
W-90.5, Ni-7.2, Fe-1.8, Co-0.45, Mo-0.05 (Ni:Fe = 4:1)	17.2	17.3	950	21	319 ± 12
W-74.0, Ni-8.0, Fe-2.0, Mo-16 (Ni:Fe = 4:1)	15.2	15.3	977	6	315 ± 18

Addition of little amounts of strengthening elements like Co, Mo improved the strength as well as elongation in these heavy alloys. Refinement of tungsten grain size due to Mo addition was achieved and this has been substantiated by other authors [7]. The elongation value of W-90.5, Ni-7.2, Fe-1.8, Co-0.45, Mo-0.05 alloy is highest (31%). This may be due to larger volume fraction of matrix (~11–15 vloume%) (Refer Fig. 18.3) present in between tungsten particles and tungsten particles are not contiguous [2]. In general, along the expected lines, the density values of W-Ni-Cu alloys were found to be more than those of W-Ni-Fe alloys.

Fractography: Tungsten heavy alloys can be considered as composites where tungsten particles are embedded in a matrix (as explained in Fig. 18.4).

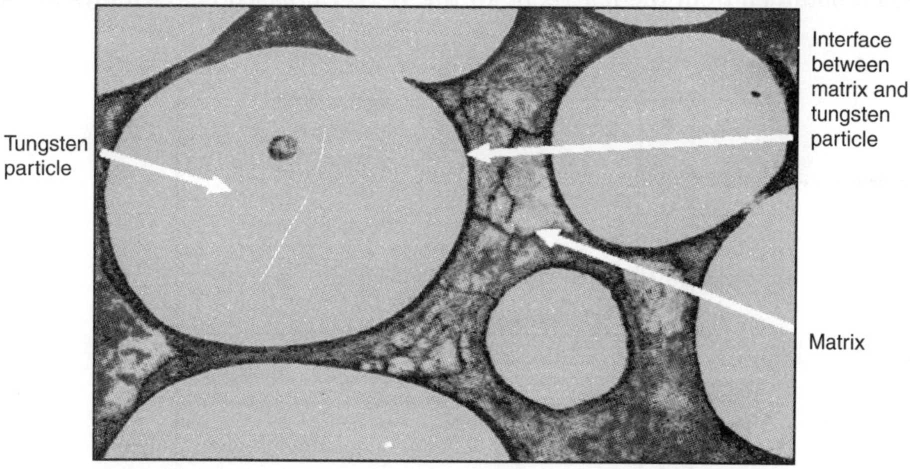

(i) Weak Matrix and/or Weak Interface : Tensile fracture by failure of matrix and/or interface (ii) Strong Matrix & Strong Interface : Tensile fracture by cleavage failure of tungsten grains

Fig. 18.4: Tungsten heavy alloy—Microstructural constituents.

If interface between tungsten particles and the matrix is weak, fracture during tensile testing would originate by interface failure. However, if the interface is strong, matrix failure may take place during tensile testing. Generally, interface failure or matrix failure is associated with the lower strength as compared to the same obtained during the cleavage fracture of tungsten particle in stronger W-alloys [8]. Figure 18.5 shows that fracture in all W-Ni-Cu base alloys occurred due to interface failure and/or matrix failure.

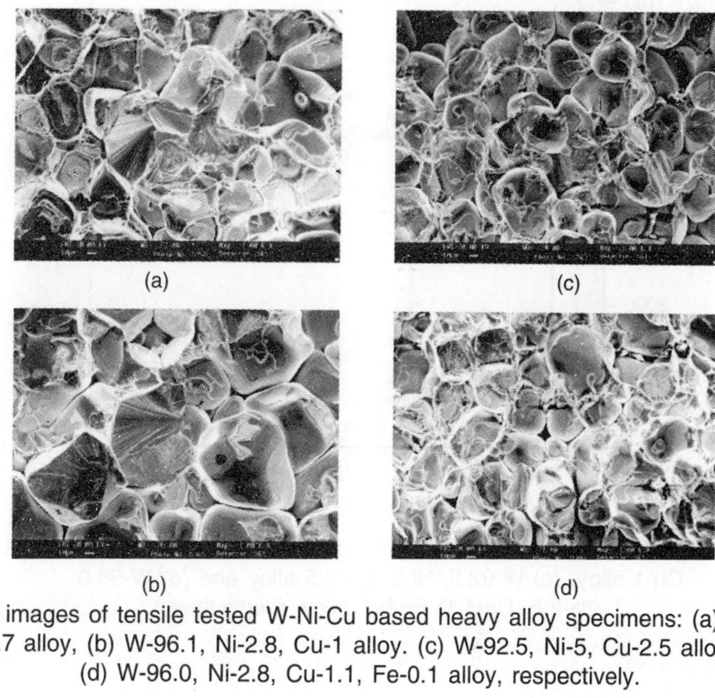

Fig. 18.5: SEM images of tensile tested W-Ni-Cu based heavy alloy specimens: (a) W-94.9, Ni-3.4, Cu-1.7 alloy, (b) W-96.1, Ni-2.8, Cu-1 alloy. (c) W-92.5, Ni-5, Cu-2.5 alloy and (d) W-96.0, Ni-2.8, Cu-1.1, Fe-0.1 alloy, respectively.

EDS spectrum obtained from the matrix of all the W-Ni-Cu based heavy alloys are shown in Fig. 18.6.

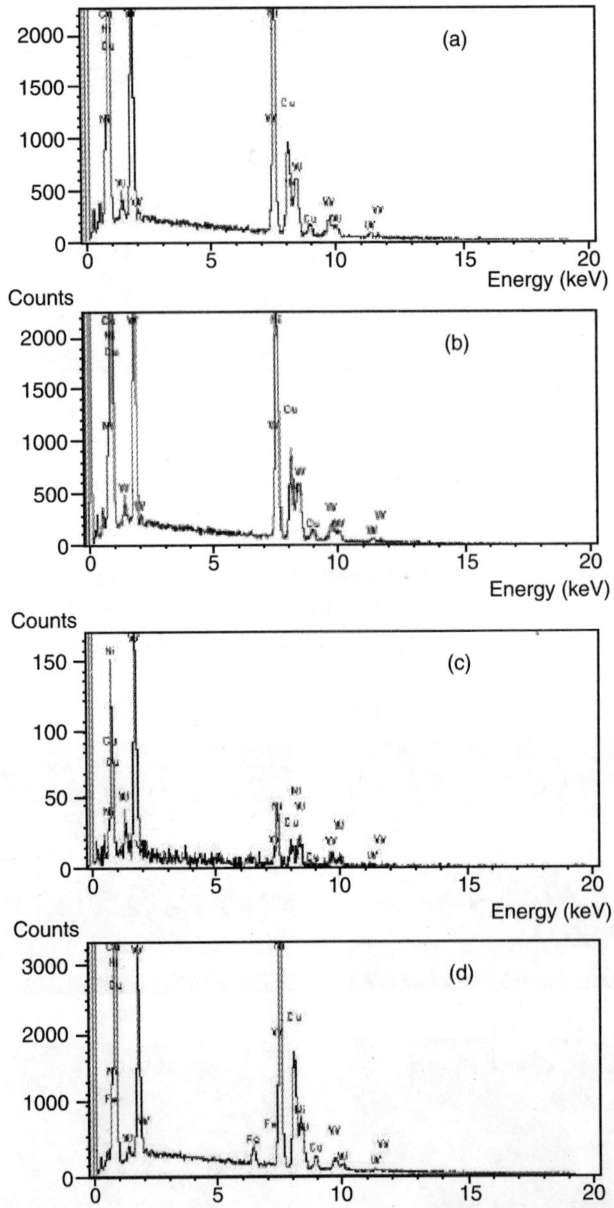

Fig. 18.6: The EDS graph obtained from matrix of (a) 94.9, Ni-3.4, Cu-1.7 alloy, (b) W-96.1, Ni-2.8, Cu-1 alloy, (c) W-92.5, Ni-5, Cu-2.5 alloy and (d) W-96.0, Ni-2.8, Cu-1.1, Fe-0.1 alloy, respectively

The EDS spectrum shows that substantial amount of tungsten is also present in the matrix. Binary phase diagram of Ni-W shows that about 29 wt% of tungsten is soluble in nickel although nickel solubility in tungsten is less than 0.3 wt% at room temperature and copper does not dissolve tungsten. In the EDS spectrum matrix shows W peaks along with Cu and Ni, because of the dissolution of some amount of tungsten by nickel. It is observed in this investigation that, interface failure and matrix failure are associated with the strength levels of up to 700 MPa. On the other hand, cleavage fracture of tungsten particle (transgranular) is associated with the strength level exceed 850 MPa. The same is observed in all W-Ni-Fe based heavy alloys. The SEM images of tensile fracture surfaces of all W-Ni-Cu based alloys are shown in Fig. 18.7.

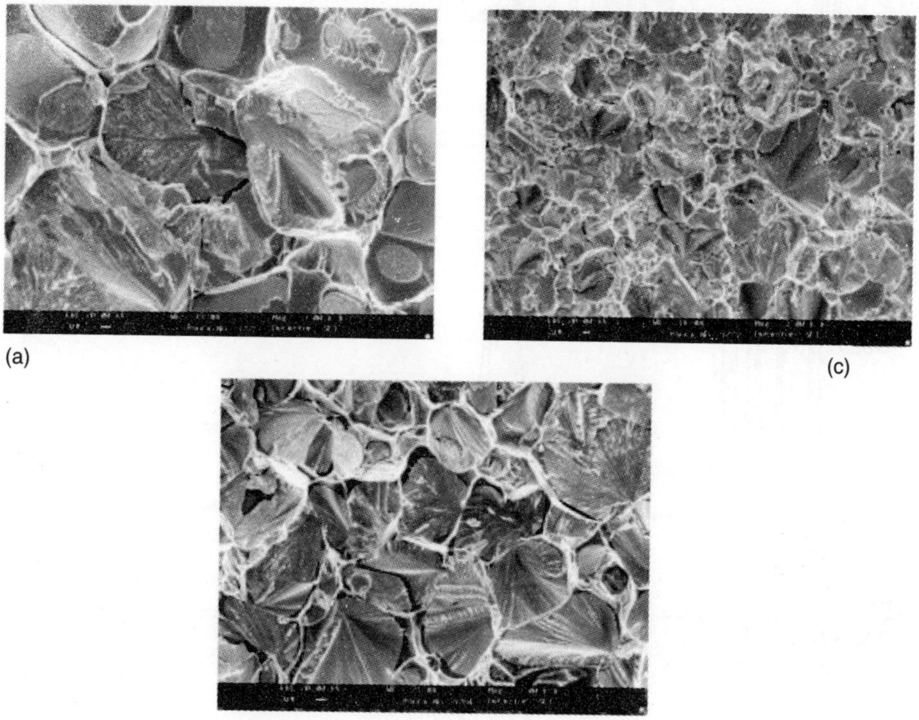

(a) (c) (b)

Fig. 18.7: SEM images of tensile tested W-Ni-Cu based heavy alloy specimens: (a) W-93.5, Ni-4.0, Fe 2.0, Co-0.5 alloy, (b) W-90.5, Ni-7.2, Fe-1.8, Co-0.45, Mo-0.5 alloy, and (c) W-74.0, Ni-8.0, Fe-2.0, Mo-16 alloy, respectively

The EDS spectrum obtained from the matrix of these alloys are given in Fig. 18.8. The EDS spectrum shows presence of substantial amount of tungsten in the matrix along with Ni, Fe, Co, Mo etc. This is due to high solubility of W in Ni. Equilibrium solubility of W in Fe and Co at room temperature is negligible, whereas Mo and W are completely soluble in each other. In this investigation it is observed that tungsten grain size plays an important role in dictating the failure mode during tensile testing. If tungsten particle size is large [refer Fig. 18.5(b)] interface failure is predominant. However, as the grain size decreases,

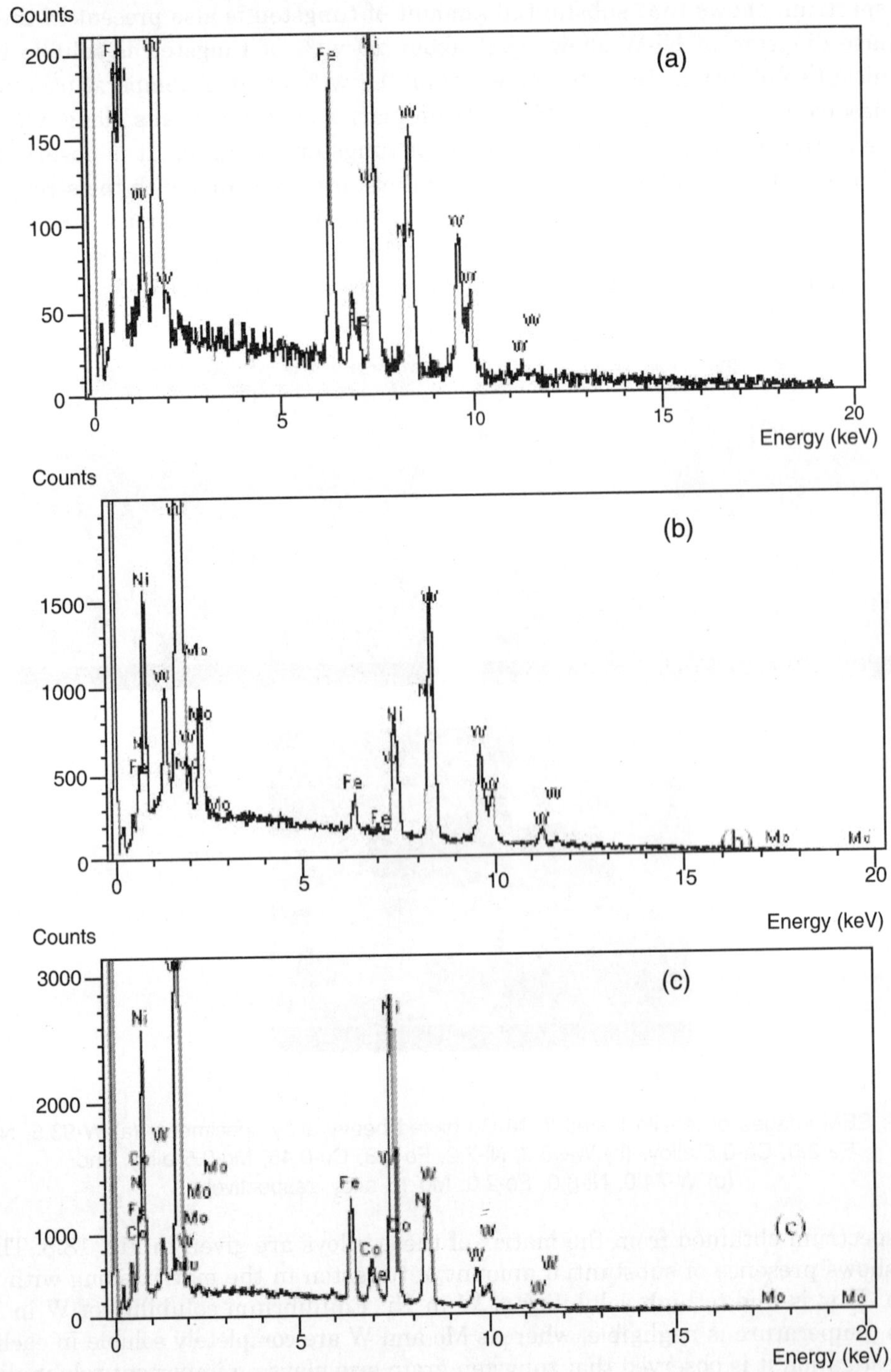

Fig. 18.8: The EDS graph obtained from the matrix of (a) W-93. 5, Ni-4.0, Fe-2.0, Co-0.5 alloy, (b) W-90.5, Ni-7.2, Fe-1.8, Co-0.45, Mo-0.05 alloy and W-74.0, Ni-8.0, Fe-2.0, Mo-16 alloy, respectively.

fracture mode changes gradually from interface failure to matrix failure and then to tungsten grain failure. Changing of fracture mode (in tensile testing) with grain size is schematically explained in Fig. 18.9.

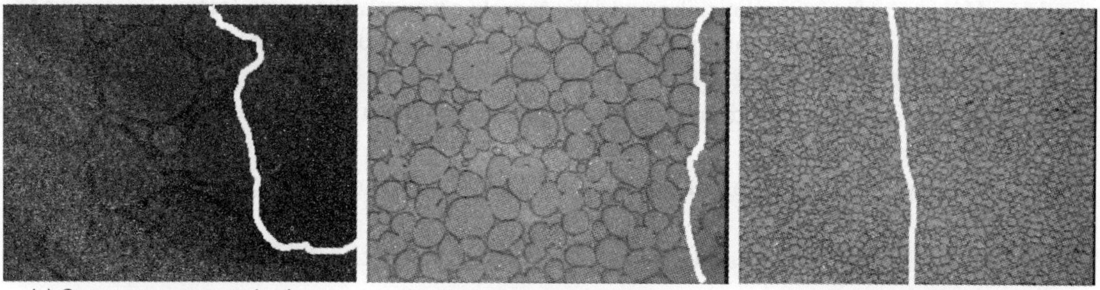

(a) Coarser tungsten grain size: intergranular fracture (predominantly interfacial failure)

(b) Medium tungsten grain size: intergranular fracture (predominantly matrix failure)

(c) Finer tungsten grain size: transgranular fracture (predominantly cleavage failure of tungsten grain)

Fig. 18.9: Sehematic diagram showing the preferred path of fracture in W-heavy alloys under tensile loading.

Specimens break predominantly by interface failure (associated with lower fracture strength), if grain size is relatively higher. For a medium grain size alloys, matrix failure is found to be predominant. If grain size is less, cleavage fracture of tungsten particles is the dominating one. This is because, in case alloys with small grain size, the energy consumption due to creation of two new surfaces will be exceedingly high (due to tortuous path) as compared to the release of interfacial energy. Therefore, cracks propagate straight by breaking tungsten particles (transgranular way) [refer Fig. 18.9(c)]. On the other hand, if grain size is larger, cracks will propagate in intergranular mode.This is because (i) intergranular crack propagation is a less energy consuming process than that of transgranular crack propagation and (ii) intergranular crack propagation path is less twisted in case of larger tungsten grain size than that in case of smaller tungsten grain size. Generally, it is apparent that the interfacial energy (between tungsten and matrix) is much lower in case of larger tungsten grain as compared to smaller tungsten grain. For a medium grain size, crack path tortuosity will not be too high to cause breaking of tungsten particle and interfacial energy will not be too low to cause interfacial failure. Hence in the medium grain size materials, the cracks propagate straight though the matrix.

It is also observed in this study that, in case of W-74.0, Ni-8.0, Fe-2.0, Mo-16 alloy, grain size is very fine. The refinement of microstructure due to Mo addition is reported in the literature [7]. However, unlike the others observation[7], this alloy does not show brittleness despite the presence of higher amount of Mo in this alloy. This alloy shows very high tensile strength and transgranular fracture was observed in the tensile tested specimens.

Quantitative relationship between tungsten grain size and the mode of fracture is not established in this study. Further studies are in progress in order to establish (i) measurement of hardness of matrix as well as tungsten particles by nano-indentation technique and (ii) fracture mode in different alloys with varied grain size. Results from these two studies will be employed to obtain quantitative relationship with the help of micro-mechanical modelling

18.4 CONCLUSIONS

1. W-Ni-Cu based heavy alloys exhibited higher density than that of the W-Ni-Fe based heavy alloys. However, former alloys showed less tensile strength than the latter. Lowering the tungsten content and increasing both the Ni/Fe ratio (up to 4) and Ni/Cu ratio (up to 2.5) resulted in increased strength of these heavy alloys.
2. W-Ni-Cu alloys are fractured predominantly through matrix failure and/or interface failure during tensile testing.
3. W-Ni-Fe alloys are broken predominantly by cleavage failure of tungsten grains in tensile testing.
4. Alloying additions like Mo and Co are observed to strengthen the W-Ni-Fe based heavy alloys.
5. Lower is the size of tungsten grain, higher is the tensile strength of these heavy alloys. As the grain size decreased, the tensile fracture mode was found to change from interface failure to matrix failure and then to tungsten grain cleavage failure.
6. Generally, cleavage failure is associated with higher tensile strength (>850 MPa) where as matrix failure/interface failure is associated with the strength level of ~700 MPa.

Acknowledgements

Authors convey their sincere gratitude to Dr. G. Malakondaiah, Director, DMRL for encouragement and kind permission to publish this work. Authors gratefully acknowledge the financial support provided by DRDO. They would also like to thank the staff members of Powder Metallurgy Group, DMRL for their help in experimentation. Authors would also like to thanks B Gangadhar Sastry and Sheo Muni Gupta Sastry for their help in microscopic examination of specimens. One of the authors (JD) is indebted to Mr. A Saibabu, for his helpful suggestions and guidance to carry out this investigation.

REFERENCES

1. Yu Yang, Hu Lianxi and Wang erde, "Microstructure and mechanical properties of a hot-hydrostatically extruded 93W-4.9Ni-2.1Fe alloy", Materials Science and Engineering A 435-436 (2006), pp.620–624.
2. Islam S H, Qu X H, Akhtar F, Feng P Z, He X B, " Microstructure and tensile properties of tungsten heavy alloys", Materials Science Forum, Vol. 534–36(1) (2007), pp. 561–564.
3. Information available at Refractory Materials & Ceramics Branch, Advanced Technology & Materials Co. Ltd., No. 76, Xueyuan, Nanlu, Haidian, Beijing 100081, China.
Email: refractory@refractorymetal.net Downloaded from www.refractorymetal.net on June 2, 2008.
4. Song, Heung-Sub Kim, Eun-Pyo Lee, Seong, "The effect of Ni/Fe ratio on the mechanical properties for tungsten heavy alloy", From Advanced Materials, Manufacturing and testing analysis centre (1998) Downloaded from http://ammtiac.alionscience.com/ammt/iacdocs.do?M005374 on 29-May, 2008.
5. Animesh Bose and Deepak Kapoor, "Processing Strategies for Tungsten Heavy alloys" Tungsten-Refractory Metals and alloys 4, Proceedings of the fourth International conference on Tungsten Refractory Metals and Alloys, Processing, Properties and Applications–1998 Edited by Animesh Bose and Robert J Dowding.

6. Boris Katavik and Milutin Nikacevic, "Properties of the Cold Swaged and Strain Aged P/M 91W-6Ni-3Co Heavy alloy". Second International conference on: Deformation Processing and Structure of Materials 26–28 May (2005), Belgrade, Serbia and Montenegro, pp. 135–140.
7. Liu Wensheng, Ma Yunzhu and Huang Baiyun, "Influence of minor element additions on microstructure and properties of 93W-4.9Ni-2.1Fe alloys", Bulletin of Materials Science, Vol. 31 (1) (2008) pp. 1–6.
8. Zhao-Hui Zhang , Fu-Chi Wang, Shu-Kui Li and Lu Wang, "Deformation characteristics of the 93W-4.9Ni-2.1Fe tungsten heavy alloy deformaed by hydrostatic extrusion", Materials Science and Engineering A 435-436 (2006), pp.632–637.
9. Jiten Das, A Chakraborty, T P Bagchi and Bijoy Sarma, "Improvement of machinability of tungsten by copper infiltration technique", International Journal of Refractory Metal and Hard Materials, In press, doi:10.1016/j.ijrmhm.2007.12.005.

❍❍

19

Development and Fabrication of U-Mo Alloy Fuel Granules for Dispersion Fuel Application in Research and Test Reactors

V.P. Sinha[a], P.V. Hegde[a], G.J. Prasad[a], G.K. Dey[b], S. Pal[a], G.P. Mishra[a] and H.S. Kamath[a]*
[a]Metallic Fuels Division; Nuclear Fuels Groups; Bhabha Atomic Research Centre;
[b]Material Science Division; Bhabha Atomic Research Centre; Mumbai, India.

Abstract

Most of the Research and Test reactors worldwide have undergone core conversion from High Enriched Uranium (HEU) base fuel to Low Enriched Uranium (LEU) base fuel under the Reduced Enrichment for Research and Test Reactor (RERTR) programme, which was launched in the late 70's to reduce the risk of nuclear proliferation. It was a challenging task for fuel designers to convert the core to LEU one without making much alteration in the reactor core design. High density uranium compounds/alloys were identified as fuel material for this purpose since higher fissile atom density can be achieved with these compounds/alloys without increasing the enrichment. In Metallic Fuels Division of BARC, R & D efforts are on to develop these high density uranium compounds/alloys. The paper describes the various fabrication methods tried for making the fuel granules of U-Mo alloy. U-Mo alloys fabricated by two different routes (*i.e.*, powder metallurgy route and melting casting route) were used to generate U-Mo alloy granules. U-10wt%Mo and U-7wt%Mo alloy samples with metastable γ-phase were given the heat treatment to get desired microstructure which shall facilitate chemical reaction between U-Mo alloys and hydrogen. The heat treated U-10wt%Mo and U-7wt%Mo alloys prepared by both the methods were then analyzed by X-ray diffraction (XRD) for phase contents. The microstructure of alloys of both compositions was determined and evaluated.

19.1 INTRODUCTION

Till mid 70's most of the research and test reactors world wide were using HEU (>85%U^{235}) based plate type dispersion fuel elements generally consisting of UAl_x (mainly UAl_3) or U_3O_8 dispersed in aluminium matrix and clad in an Al-alloy [1]. In the late 70's the RERTR programme was initiated to address the concern related to proliferation and diversion issues. The main aim for most of the fuel designers was to develop a fuel cycle which is based on LEU (< 20%U^{235}) rather than HEU. This type of development was also required because many of the researchers wanted to develop a neutron source with high thermal neutron flux density for superior quality experiments. This can be achieved either by increasing the fuel loading or by using higher uranium density compounds/alloys. The most common fabrication techniques involving LEU based dispersion fuel are picture framing followed by roll bonding technique for plate fuel and vibrocompaction followed by infiltration or extrusion for rod type or tube type fuel. Plate type fuel geometry by roll bonding method does not allow the fuel particle volume to exceed 55% and in the case of rod type and tube type geometry the maximum fuel particle volume loading is still lower [2-3]. Due to the imposed limitation on the increase of fuel particle volume in the fuel meat it was essential to increase the heavy metal density (*i.e.*, Uranium) in the fuel itself. There are two ways of doing it, *i.e.*, one is to increase the uranium metal density by using uranium metal itself and the other one is by using high density intermetallics/ compounds of uranium.

There are several intermetallics/compounds of uranium which were considered because of their higher heavy metal density (see Table 19.1). The U_3Si compound and intermetallics of U_6Me where Me can be Fe, Ni, Mn or Ge gives higher uranium metal density but their high swelling rate under low burn up during irradiation is major road block. The studies show their poor irradiation behaviour for thin plate type dispersion fuel geometry due to the breakaway swelling under relatively low burn ups [4-6]. On the other hand, U_3Si_2 dispersion fuel in aluminium matrix clad in Al-alloy shows stable irradiation behaviour, but the maximum uranium metal density achieved in the fuel meat is in the range from 4.8 to 6.0 gm/cm^3 [7]. One way is to use uranium metal dispersoid for plate or rod type geometry fuel elements to achieve the uranium metal density of the order of 8 to 9.0 gm/cm^3 in the fuel meat.

The room temperature phase of pure Uranium (i.e. α-uranium having orthorhombic crystal structure) shows poor irradiation behaviour in the form of thermal and irradiational growth during reactor operation which makes it unsuitable as fuel at higher reactor operating temperatures. The high temperature phase of uranium (*i.e.*, γ-uranium having body centered cubic crystal structure) is more stable under irradiation in than the α-phase. Thermodynamically the γ-phase is not stable during fabrication and irradiation conditions, however there are some alloying elements which make this cubic phase metastable for an indefinitely long period at room temperature and for a long time period at elevated temperature. Several transition metals, particularly from Group IVA to Group VIIIA of periodic table form solid solution with γ-U, which can be retained as metastable phase at room temperature upon cooling. The stabilizing power of these elements increases with their atomic number since more and more d-electrons will participate in bonding due to hybridization with *s* and *p* orbital electrons. However, their solubility will decrease as the size difference with uranium atom increases. In fact the increase in bond strength with the increase in atomic number of transition elements will promote the formation of intermetallic compound rather than a solid solution. The alloys which have a tendency to form stable cubic phase of uranium at room temperature are U-Cr, U-Mo, U-Nb, U-Re, U-Ru, U-Ti, U-V, U-Zr etc. The first two transition elements of 4d series (*i.e.*, Zr and Nb) will form complete solid solution with γ-uranium but to retain the 100% γ-phase of uranium at room temperature, large concentration of these elements are required to be added in the alloy. On the other extreme Pd and Pt have only ~2 at% solubility in γ-uranium and these alloying systems will give very stable intermetallic compounds with uranium. It was earlier recognized that molybdenum gives a good compromise for the two extremes mentioned above.

Table 19.1: Density of various fuel compounds and alloys of uranium

Compound	Uranium	UAl_4	UAl_3	UAl_2	USi	U_3Si_2	U_3Si	U_6Mn	U_6Fe	U_6Ni	U-2wt% Mo	U-5wt% Mo	U-10wt% Mo
Theoretical density (gm/cm^3)	19.07	6.2	6.8	8.14	10.96	12.2	15.3	17.4	17.6	17.8	18.9	18.6	18.18

In uranium molybdenum system under equilibrium cooling, the cubic γ-phase with the space group Im3m will break into orthorhombic α-phase of space group Cmcm and body centered tetragonal γ′-phase (U_2Mo) of space group I4/mmm (see Fig. 19.1). However under rapid cooling conditions the metastable γ-phase can be retained at the room temperature. The U-Mo alloys containing less than 11 at% Mo when water quenched from the gamma phase region may be called as alpha phase alloys while those with more than 11 at% Mo may be called as gamma phase alloys under the same conditions [8-12]. In fact the U-Mo alloys containing upto 10 at% Mo when water quenched forms α′ and α″ phases which are slight variation from the orthorhombic α-phase [13-14]. Tangri has reported base centered tetragonal phase γ° for water quenched alloys containing 11.39 at% to 12.73 at% Mo [15]

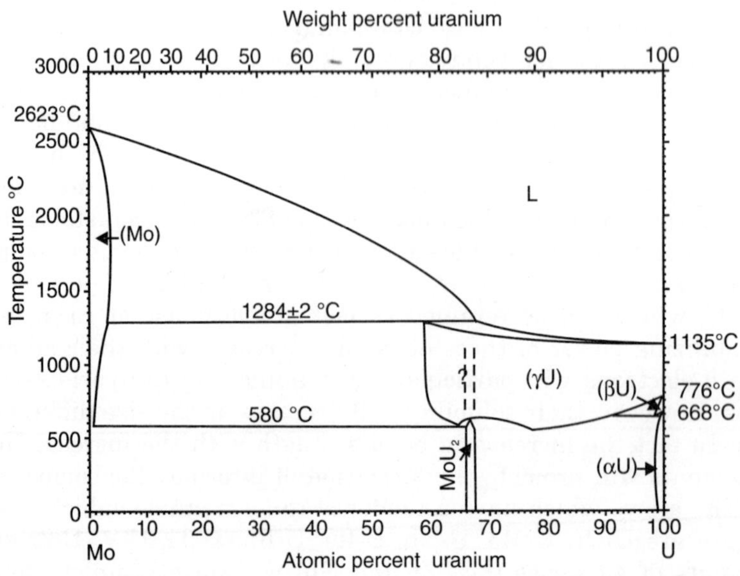

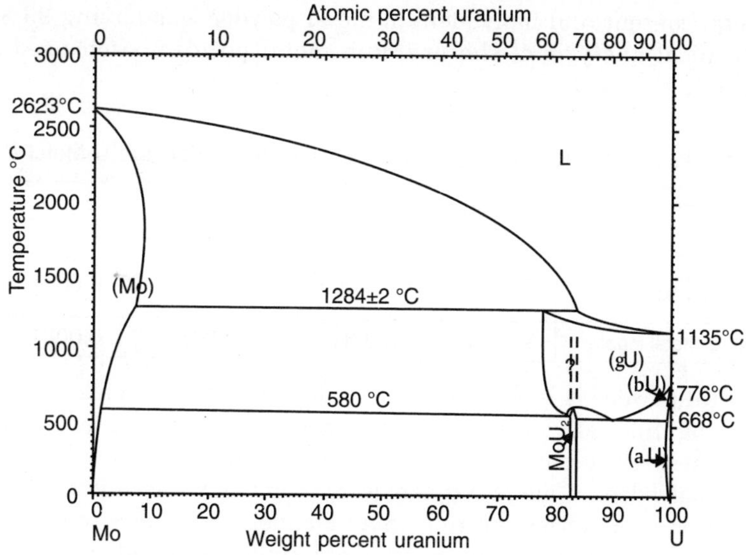

Fig. 19.1: Molybdenum uranium phase diagram

At Metallic Fuels Division, BARC efforts are on to develop high density uranium alloys/ compounds for dispersion type plate fuel development and the present study is an extension to this programme. The fuel particles (*i.e.*, dispersoids) are required to be dispersed in the non fissile matrix for dispersion fuels in research and test reactors. Brittle compounds of uranium can easily be crushed into desired size fuel granules with mechanical crushing or grinding. But this is very difficult in case of uranium alloys since the metastable γ-phase of uranium is ductile. Alternate methods are explored to generate fuel granules of required size from the ductile uranium alloys. This paper covers the different methods explored in MFD, BARC for the fabrication of U-Mo alloy fuel granules. The fabrication of U-Mo alloy was carried out by an innovative powder processing route as well as conventional ingot metallurgy route. To generate U-Mo alloy fuel granules by hydriding dehydriding method, tailored microstructure was produced by proper heat treatment of alloys. The U-Mo alloy fuel granules fabricated by powder processing route or by hydriding dehydriding of pellets is characterized by XRD method for phase identification. The XRD analysis result of heat treated U-Mo alloys fabricated by both the powder metallurgy route as well as melting casting route and features of microstructures developed for the heat treated U-Mo alloys are discussed in the paper.

19.2 EXPERIMENTAL WORK

19.2.1 Fabrication of U-10wt % Mo Alloy by Powder Metallurgy Route

19.2.1.1 Fabrication of U-10wt% Mo alloy pellet

The starting material for the purpose is uranium metal powder with average particle size of around 30 μm [shown in Fig. 19.2(a)] and molybdenum metal powder with average particle size of around 10 μm [shown in Fig. 19.2(b)]. Uranium metal powder used here was having

99.6 wt% 'U' metal content and molybdenum metal powder was having 99.8 wt% 'Mo' metal content. The chemical analysis of the uranium metal powder batch used for this work is shown in Table 19.2.

Table 19.2: Chemical characteristics of U-Metal powder and U-Metal ingot

Chemical form	Uranium (wt %)	Carbon (wt %)	Nitrogen (wt %)	Boron (wt %)
U-metal powder	99.6±0.1	0.063	0.050	0.0001
U-metal ingot	99.9±0.1	0.040	0.005	0.0001

Uranium metal powder is highly prone to oxidation hence extreme care was taken during the powder handling. Specially designed glove boxes with very good leak tightness were used for the purpose. These glove boxes were circulated with once through high purity argon gas. The uranium and molybdenum metal powders were thoroughly mixed in the required proportion. The mixing operation was carried out in a small blender for 4 hours to get good homogeneous mixture. During the mixing operation no admixed binder was added. The mixture was then compacted at 600 MPa pressure in a 20 mm die into cylindrical pellets.

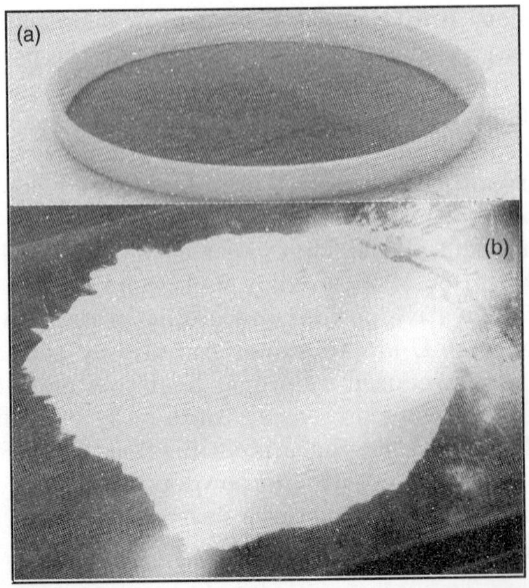

Fig. 19.2: Photograph of starting material. (a) Uranium metal powder. (b) Molybdenum metal powder.

Die wall lubrication was used during compaction to reduce die wall friction. The L/D ratio of compacted pellets was kept around 0.5. The green pellets were then loaded on molybdenum trays coated with plasma sprayed alumina. The charge carrier was then loaded inside the 10 kW resistance heated furnace and the compacts were heated at 1050±10°C for 6 hours under vacuum. The impurity pick up during high temperature synthesis was avoided by repeat evacuation and purging the furnace housing inside the glove box with high purity argon gas before starting heating operation. The furnace was allowed to cool under vacuum and the product *i.e.*, U-Mo alloy pellets were taken out (shown in Fig. 19.3) [16].

Fig. 19.3: Photograph of U-10wt%Mo alloy pellets prepared by powder metallurgy route.

19.2.1.2 Fabrication of U-10wt% Mo alloy fuel granules

U-Mo alloy pellets/billets prepared by either powder metallurgy route or melting casting route are difficult to granulate to required size with conventional jaw crusher. These alloys are ductile in nature hence do not fracture easily under the application of load at room temperature. Keeping this point in mind a novel powder processing method was tried to generate U-Mo fuel granules. In this method the uranium metal powder [shown in Fig. 19.2(a)] and molybdenum metal powder [shown in Fig. 19.2(b)] were mixed in required proportion (*i.e.*, for U-10wt% Mo composition). The mixture was then compacted at 600 MPa pressure in a 20 mm die to get green cylindrical compacts. These green compacts were then crushed mechanically and sieved to get granules in green stage. The green granules were then loaded on molybdenum trays coated with plasma sprayed alumina. The charge carrier was then loaded inside the 10 kW resistance heating furnace and the green granules were heat treated at 1050 ± 10°C for 6 hours under vacuum. Entire operation was carried out inside the glove box which was purged continuously with high purity argon gas. The furnace was allowed to cool under vacuum and the product (*i.e.*, fuel granules of U-10wt% Mo alloy) was taken out (shown in Fig. 19.4). The fuel granules of U-10wt% Mo alloy synthesized by the above method were further reduced in size by crushing in a jaw crusher. The detailed flow sheet for the process is shown in Fig. 19.5.

19.2.2 Alloy Preparation by Induction Melting Route

With a view to compare the characteristics of alloys prepared by powder metallurgy route and that of melting and casting route, the alloys were also prepared by conventional induction melting route. In case of induction melting, the cylindrical uranium metal pellets (shown in Fig. 19.6) and molybdenum metal pieces were used as starting materials for alloy preparation. Uranium metal pieces were having 99.9 wt% 'U-content' and molybdenum metal pieces were having 99.8 wt% 'Mo-content'. The uranium metal pellets used here were approximately of 12 mm diameter and around 10 mm in length. The chemical analysis of uranium metal pellets used for this work is shown in Table 19.2.

Fig. 19.4: Photograph of U-Mo alloy granules fabricated by powder metallurgy route

The uranium metal pellets were electro polished to remove surface oxidation layer and then charged in a Yitria coated graphite crucible along with molybdenum metal pieces in required proportion for melting. The crucible was then loaded in an induction heating furnace and the melting was done at 1500±10°C for 30 minutes. The furnace was then allowed to cool under vacuum and ingots were taken out (shown in Fig. 19.7).

19.2.3 Fabrication of U-Mo Alloy Fuel Granule from Pellet/Billet by Hydriding Dehydriding Method

The method of hydriding and dehydriding was also tried to generate U-Mo alloy fuel granules. Two different alloy compositions *i.e.*, U-7wt%Mo billet fabricated by melting casting route and U-10wt%Mo billet/pellet fabricated by both melting casting route and powder metallurgy routes were used for hydriding experiments. Due to the fact that hydrogen has no reaction with molybdenum it is quite difficult to generate hydride powders by reacting hydrogen with γ-uranium. The solid solubility of molybdenum in γ-uranium is high (~ 35 at%) while the same for α-uranium is very less (~1at% at room temperature) as per the phase diagram. On the other hand it is possible that hydrogen may react with α-uranium since it has very less solubility for molybdenum. As per the U-Mo phase diagram a eutectoid reaction takes place below 550°C isotherm where γ-phase will break into α-phase and U_2Mo intermetallic.

In hydriding dehydriding method, tailored microstructure was developed for U-Mo alloys by giving appropriate heat treatment. The alloys were encapsulated in quartz tube under the partial pressure of helium. The capsules were then heat treated at 500°C for 48 hours to get proeutectoid α-uranium phase along the prior g grain boundary. The proeutectoid phase (*i.e.* α-uranium) will react with hydrogen to form UH_3. Since the density of UH_3 is very less as compared to uranium metal, it falls as loose powder particle. Hydriding was tried with both the alloy compositions *i.e.*, U-7wt%Mo alloy fabricated by melting casting route and U-10wt%Mo alloy fabricated by powder metallurgy route and melting casting route. Initial hydriding runs were taken with Ar-8%H_2 gas in a resistance heating furnace with glove box housing where temperature was kept at 400°C for 8 hours. UH_3 formation with Ar-8%H_2 gas mixture was almost negligible [shown in Fig. 19.8(a)] and hence pure hydrogen was tried for both the alloy compositions outside the glove box. In this case a tubular furnace was charged with the alloy piece and pure hydrogen was allowed to flow. The temperature for one complete hydriding cycle was kept at 400°C for 2 hours. UH_3 fine particles were collected inside the quartz tube after a complete hydriding cycle [shown in Fig. 19.8(b)]. The hydride powders fabricated by this method were extremely pyrophoric and lots of precautions were taken during material transfer.

19.2.4 Characterization

U-7wt%Mo alloy fabricated by melting casting route and U-10wt%Mo alloy fabricated by both powder metallurgy route and melting casting route were then analyzed by XRD for phase determination, where scan rate was kept at 1°/minute. U-Mo alloy samples after the heat treatment were also characterized by XRD technique for phase analysis. Theta theta standard XRD machine was used to generate XRD patterns. The source was Cu Kα and the tube voltage was 35.0 kV and the tube current was 28.4 mA. Tablet of uranium molybdenum alloys prepared by powder metallurgy route were used to get XRD pattern. The XRD sample for induction melted alloys were prepared by cutting the cast ingots by silicon carbide wheel in small disk form and then polished using 600 μm emery polishing paper to get the good surface finish.

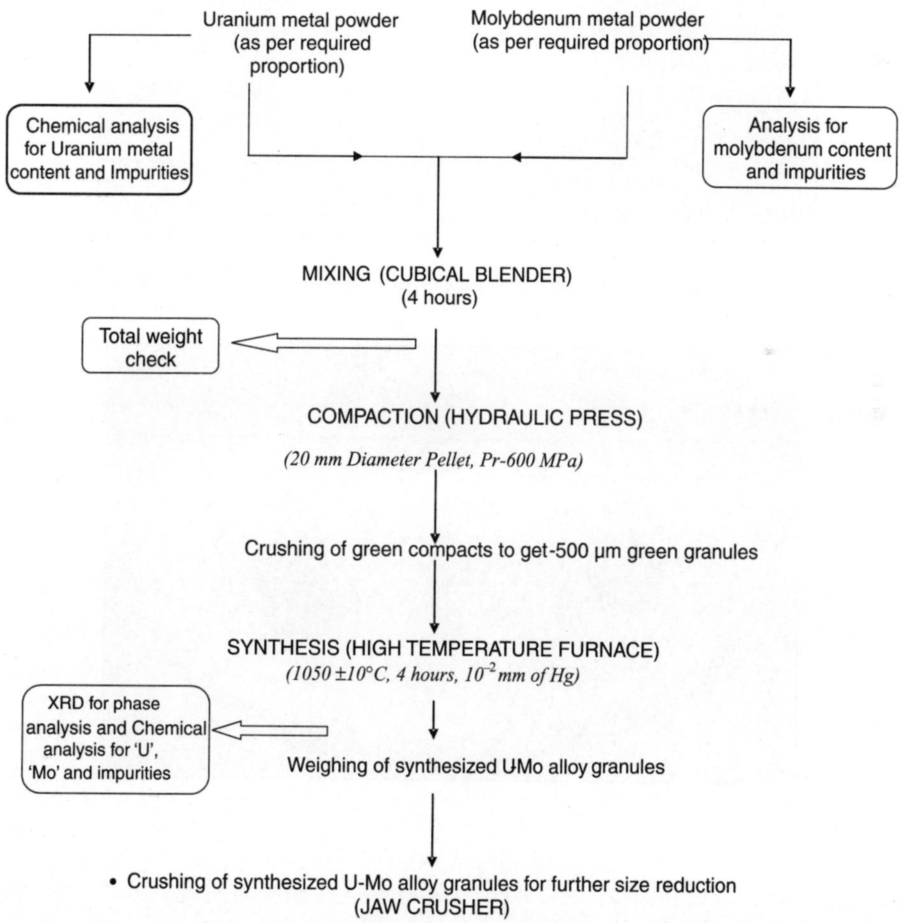

Fig. 19.5: Process flow sheet for preparation of U-Mo alloy fuel granules by PM route

Fig. 19.6: Photograph of uranium metal pellets

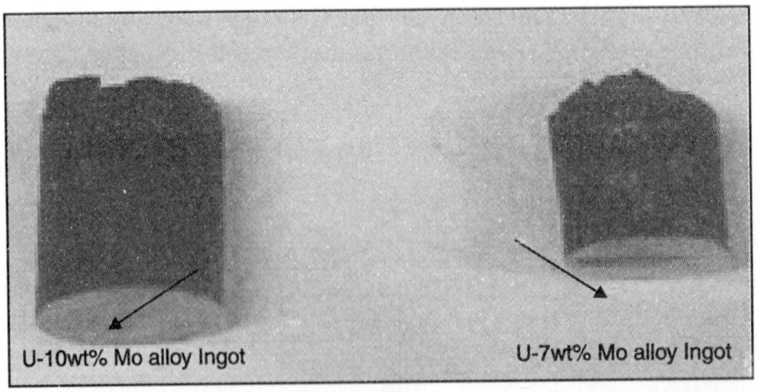

Fig. 19.7: Photograph of U-Mo alloy Ingots prepared by induction melting route

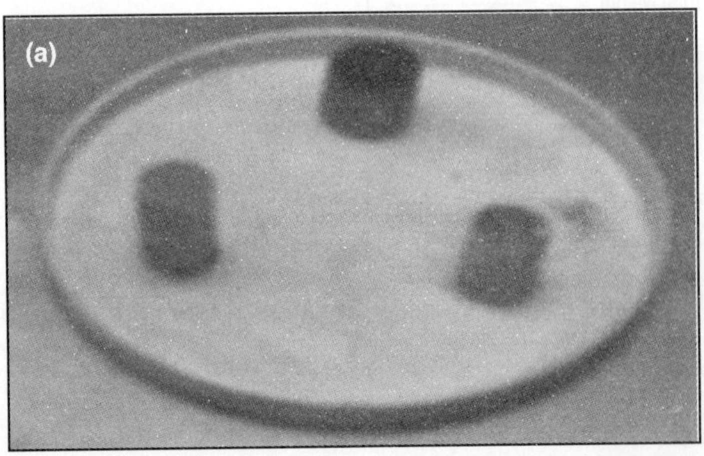

Fig. 19.8: Photograph of U-Mo alloy pellets after hydriding. (a) Hydriding with Ar-8%H_2 mixture. (b) Hydriding with pure H_2 gas (UH_3 powder generated)

Microstructure for U-Mo alloy samples with different fabrication and heat treatment history were developed and evaluated. The metallographic samples of all the alloys were prepared by cold mounting. The mounted samples were ground and polished by using standard metallographic procedures and equipment. The polished samples were etched by using freshly prepared etchants [17-18].

19.3 RESULTS AND DISCUSSION

19.3.1 Phase Analysis

The XRD pattern for U-7wt%Mo alloy prepared by induction melting route (as shown in Fig. 19.9) shows the presence of major phase as g-uranium. In case of U-10wt%Mo alloy prepared by powder metallurgy route [as shown in Fig. 19.10(a)] and by induction melting route [as shown in Fig. 19.10(b)], the XRD pattern reveals the complete g-uranium formation. Similarly the XRD pattern for heat treated U-7wt%Mo alloy [shown in Fig. 19.11(a)] shows the presence of cubic γ-uranium and also that of α-uranium.

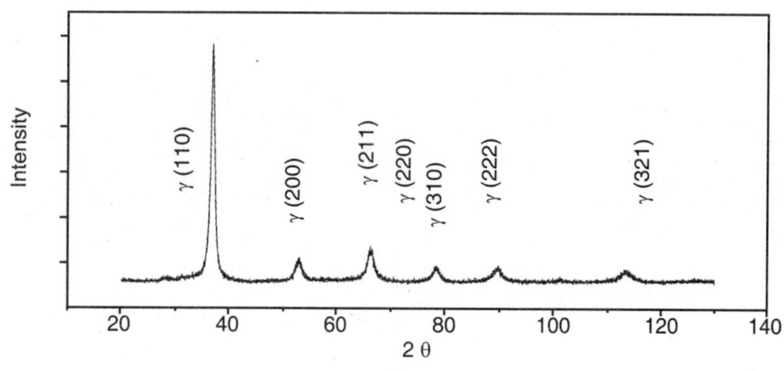

Fig. 19.9: X-Ray diffraction pattern of U-7wt% Mo alloy prepared by induction melting route

The XRD pattern of heat treated U-10wt%Mo fabricated by melting casting route [as shown in Fig. 19.11(b)] and also by powder metallurgy route [as shown in Fig. 19.11(c)] shows the presence of cubic γ-uranium as major phase content while α-uranium in small percentage. Heat treatment of U-Mo alloys with 7wt%Mo and 10wt%Mo as alloying elements was carried out at 500°C for 48 hours. As per the equilibrium phase diagram of U-Mo (shown in Fig. 19.1) below 550°C the cubic γ-uranium will go under eutectoid decomposition to form orthorhombic α-phase and intermetallic compound (U_2Mo) with body centered tetragonal structure. The XRD pattern for heat treated U-7wt%Mo and U-10wt%Mo alloys does not show any U_2Mo phase as the reaction product. The reason for not detecting any amount of U_2Mo may be either the phase is in such a small quantity that XRD was not able to detect it or local molybdenum concentration is less than 18 wt% (which is required to U_2Mo formation) as the diffusion of molybdenum is very slow at these temperatures. To get U_2Mo as one of the reaction product the alloys may have to be heated and kept for longer duration below the eutectoid decomposition temperature.

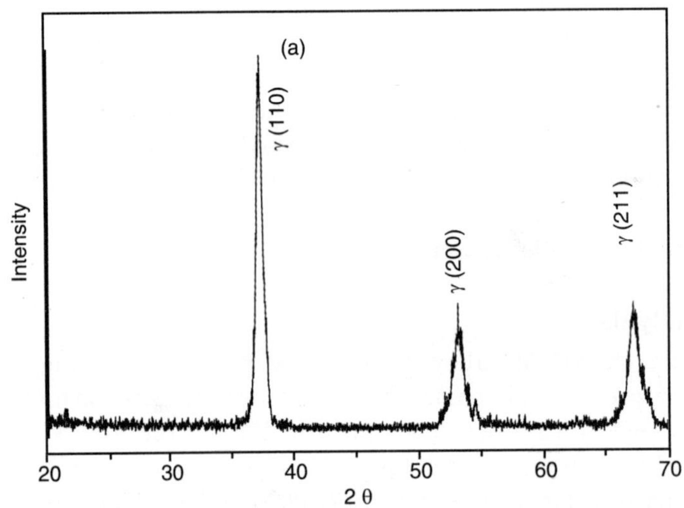

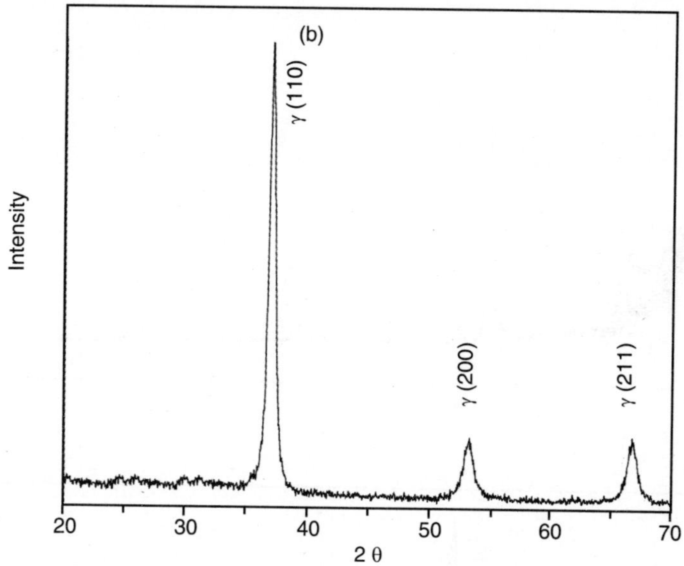

Fig. 19.10: X-Ray diffraction pattern of U-10wt%Mo alloy pellet prepared by (a) Powder metallurgy route, (b) Induction melting route

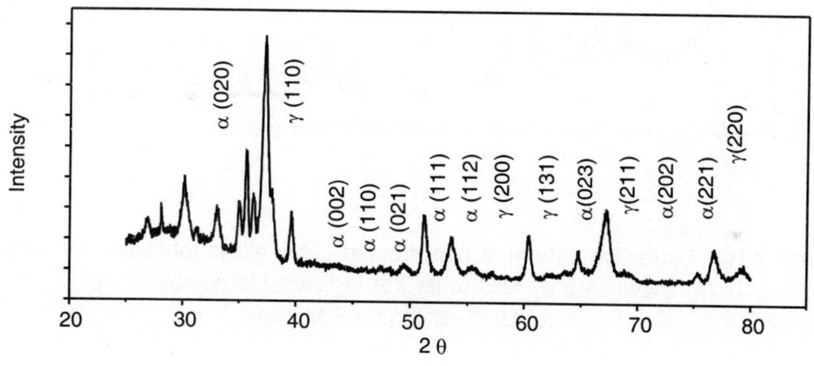

(a)

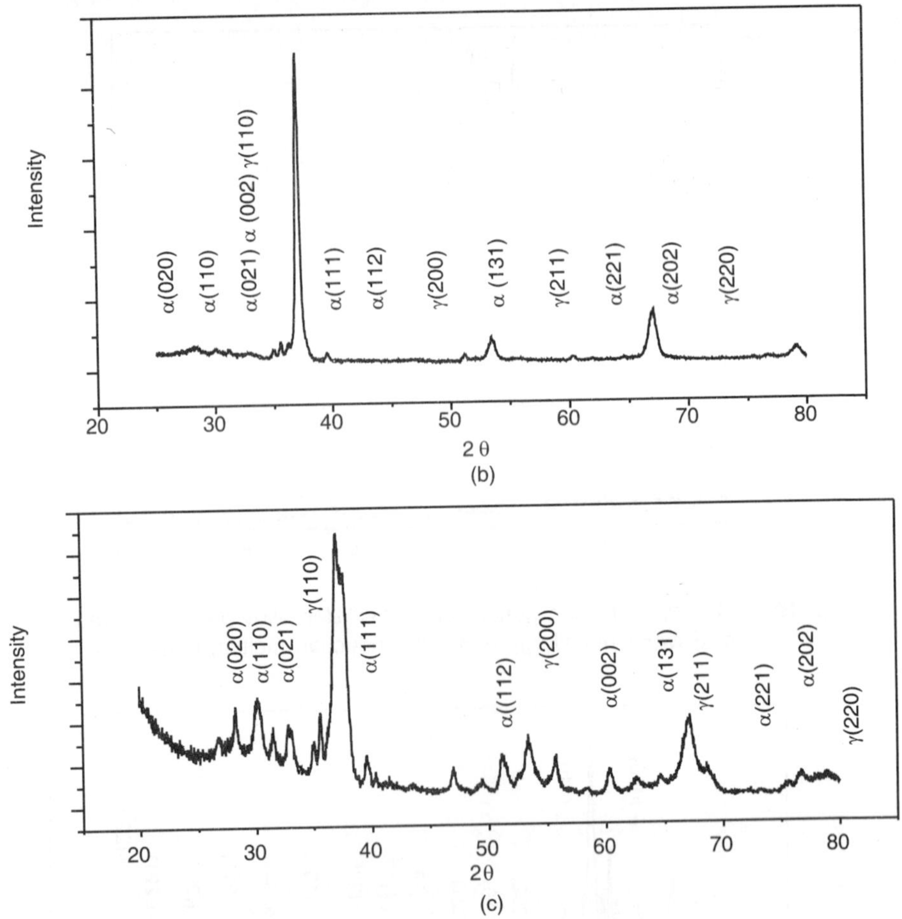

Fig. 19.11: X-Ray Diffraction pattern of heat treated U-Mo alloys fabricated by various routes (a) U-7wt%Mo by M/C route, (b) U-10wt%Mo by M/C route, (c) U-10wt%Mo by P/M route

19.3.2 Optical Microscopy

The microstructure of U-7wt%Mo alloy fabricated by melting casting route (as shown in Fig. 19.12) shows an average grain size γ-uranium of around 60 μm. The microstructure also shows the presence of second phase along the prior γ- grain boundary. The microstructure of U-10wt% Mo alloy prepared by powder metallurgy route [as sown in Fig. 19.13(a)] and by induction melting route [as shown in Fig. 19.13(b)] shows single phase grains of γ-uranium with an average grain size of around 50 mm. The microstructure of U-Mo alloys prepared by powder metallurgy route shows some dark regions. These are the pull outs which took place during the grinding and polishing operations because the alloys prepared by powder metallurgy route were slightly porous in nature and the density was around 85%T.D. The microstructure of heat treated U-7wt% Mo alloy fabricated by melting casting route [as shown in Fig. 19.14(a)] and U-10wt% Mo alloy fabricated by melting casting route [as shown in Fig. 19.14 (b)] and by powder metallurgy route [as shown in Fig. 19.14 (c)] shows the formation of proeutectoid α-uranium phase along the prior α-grain boundary. The lamellar decomposition can be seen in the proeutectoid phase which suggests nucleation and growth as the reaction mechanism. Most of the α-uranium grain has remained untransformed during the heat treatment which can be seen as bright regions in the microstructures.

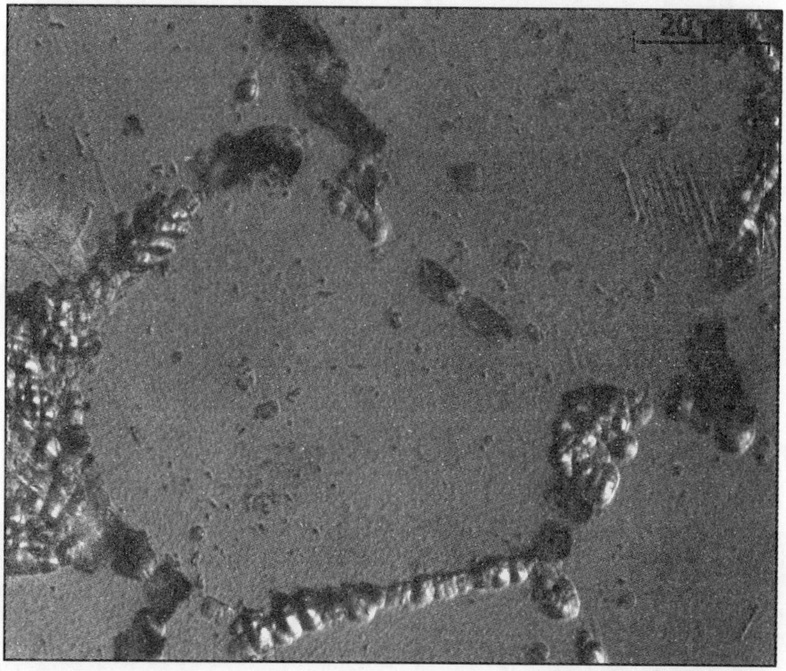

Fig. 19.12: Optical photomicrograph of U-7wt%Mo alloy prepared by induction melting route

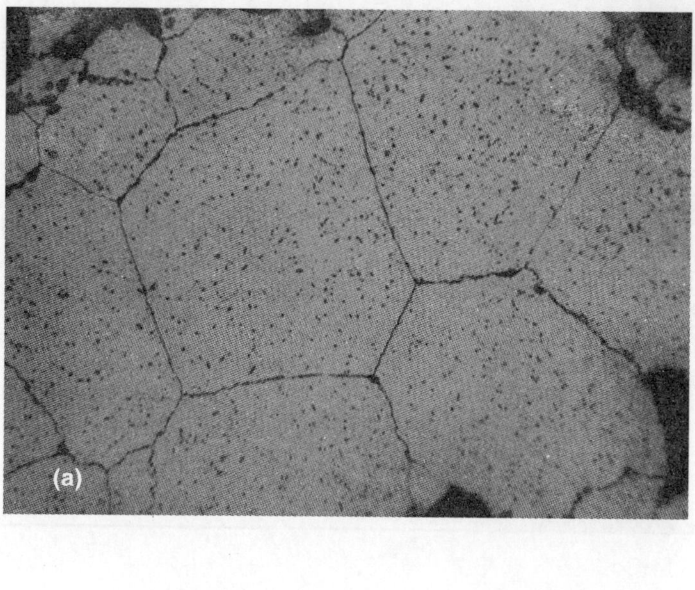

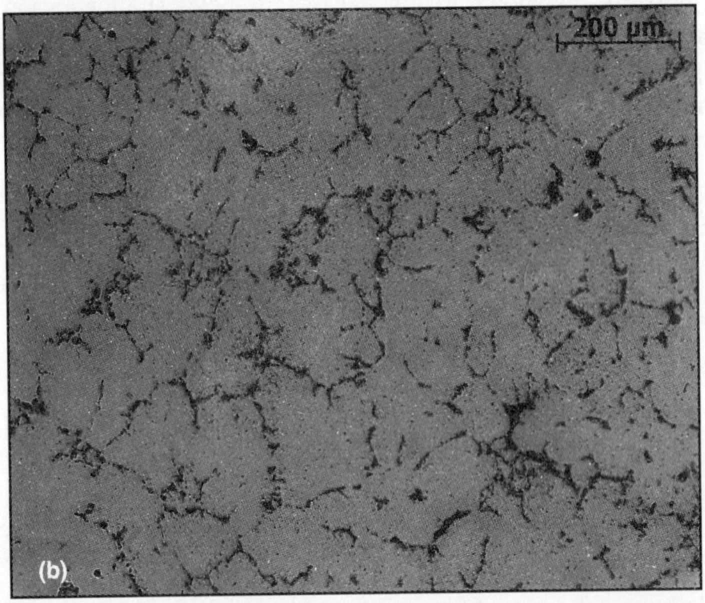

Fig. 19.13: Optical photomicrograph of U-10wt%Mo alloy pellet prepared by (a) Powder metallurgy route, (b) Induction melting route.

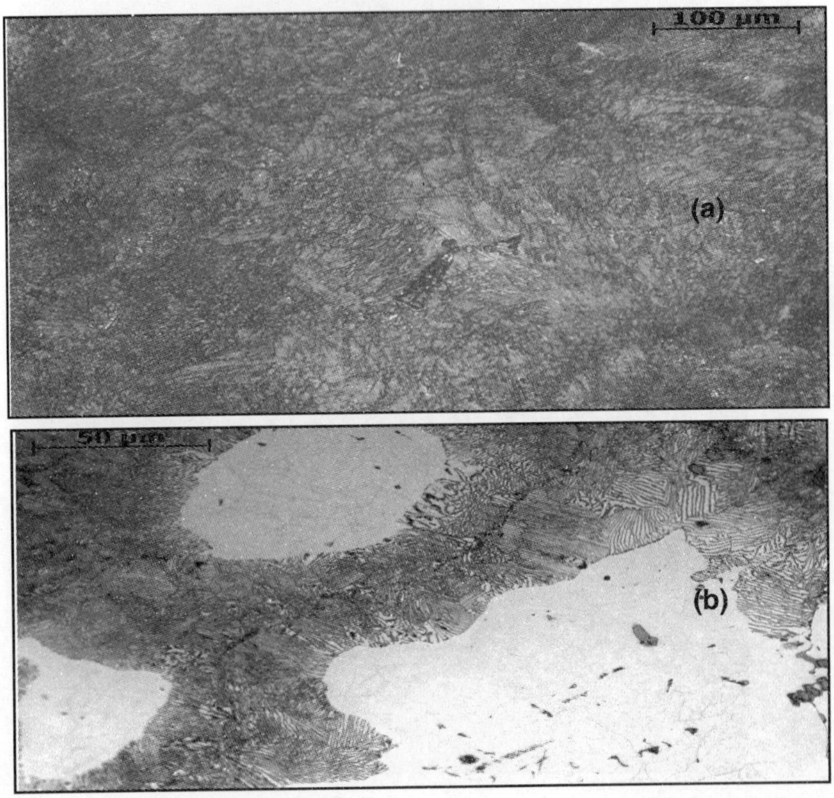

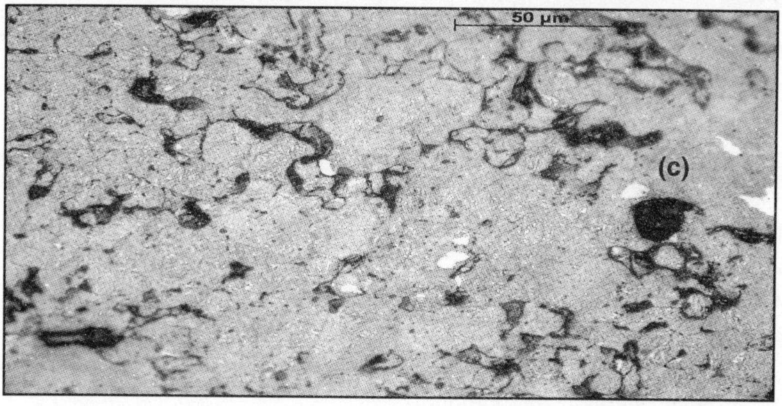

Fig. 19.14: Optical photomicrograph of heat treated U-Mo alloys fabricated by various routes (a) U-7wt%Mo by M/C route, (b) U-10wt%Mo by M/C route, (c) U-10wt%Mo by P/M route.

19.4 CONCLUSIONS

The uranium molybdenum alloy pellets prepared by powder metallurgy method had shown excellent integrity and no lamellar or radial cracks were observed with the naked eye. The alloys prepared by powder metallurgy route have shown very similar phase and microstructural characteristics as that of alloys prepared by induction melting route for same composition. U-Mo alloy fuel granules were successfully prepared by an innovative powder metallurgy route for U-10wt%Mo composition. The XRD result shows the presence of metastable high temperature cubic γ-uranium phase and room temperature orthorhombic α-uranium phase. The product fabricated (*i.e.*, U-10wt%Mo alloy fuel granules) was not single phase hence the optimization of process parameters are required to fabricate the U-Mo fuel granules in single cubic γ-uranium phase. Hydriding of U-Mo alloys with Ar-8%H_2 mixture do not yield any UH_3 particles while the yield with pure H_2 was around 18 to 20%. Hence more improvement in tailored microstructure and optimization of some process parameters is required to get higher yield during hydriding.

Acknowledgements

The authors are grateful to Dr. (Mrs) S.B. Roy Head, Uranium Extraction Division for the supply of Uranium metal powder and Mr. V.D. Pandey for microstructural examination of samples. The authors also deeply appreciate the help and cooperation of Mr. Arun Kumar Head, Radio Metallurgy Division and Mr. R.P. Singh, Associate Director, Nuclear Fuels Group.

REFERENCES

1. R.F. Domagla, T.C. Wiencek and H.R. Tresh, Nucl. Tech. 62 (1983)353.
2. J.P. Durand, Y. Lavastre and M. Grasse, Proceedings of the 20th International Meeting on Reduced Enrichment for Research and Test Reactors, Argonne National Laboratory Report ANL/TD/TM99-06(1999)28.
3. H.B. Peacock, Coextrusion of 60 to 30 wt% U3O8 Nuclear Fuel Elements, DPST-80-447, E.I. du Pont de Nemours and Company (1980).
4. M. Ugajin, A. Itoh, M. Akabori, N. Ooka and Y. Nakakura, Irradiation behaviour of high uranium-density alloys in the plate fuels, J. Nucl. Mater. 254(1998)78.
5. M.K. Meyer, T.C. Wiencek, S.L. Hayes and G.L. Hofman, Irradiation behaviour of U_6Mn-Al dispersion fuel elements, J. Nucl. Mater. 278(2000)358.
6. G.L. Hofman, R.F. Domagala and G.L. Copel, Irradiation behaviour of low-enriched U_6Fe-Al dispersion fuel elements, J. Nucl. Mater. 150(1987)238.
7. G.L. Copeland *et al.*, Performance of Low Enrichement U_3Si_2-Aluminium dispersion fuel elements in the Oak Ridge Research Reactors, Argonne National Laboratory Report, ANL/RERTR/TM-10 (1987).
8. Ki Hwan Kim *et al.*, Characterization of U-2wt% Mo and U-10wt% Mo alloy powders prepared by centrifugal atomization, J. Nucl. Mater. 245(1997)179.
9. M.K. Meyer *et al.*, Low Temperature Irradiation behaviour of uranium-molybdenum alloy dispersion fuel, J. Nucl. Mater. 304(2002)221.
10. S. Nicholson, D.G. Harris and J.J. Stobo, The effect of ternary additions on the microstructures of dilute U-Mo alloys, J. Nucl. Mater. 12(1964)173.
11. R.L. Craik, D. Birch, C. Fizzotti and F. Saraceno, Phase equilibria in Uranium-rich binary alloys containing molybdenum and zirconium and the effect of ternary additions of carbon, J. Nucl. Mater. 6(1962)13.
12. Gerard L. Hofman and Mitchell K. Meyer, Design of high density gamma-phase uranium alloys for LEU dispersion fuel applications, RERTR conference, Sao Paulo, Brazil, October 18–23, (1998).
13 C.R. Clark and M.K. Meyer, Fuel powder production from ductile uranium alloys, RERTR conference, Sao Paulo, Brazil, October 18–23, (1998).
14 J.P. Durand, J.C. Ottone, M. Mahhe and G. Ferraz, LEU fuel development at CERCA status as of October 1998, RERTR conference, Sao Paulo, Brazil, October 18–23, (1998).
15. K. Tangri and G.I. Williams, Metastable phases in the uranium-molybdenum system and their origin, J. Nucl. Mater. 4(1961)226.
16. V.P. Sinha, G.J. Prasad, P.V. Hegde, R. Keswani, C.B. Basak, S. Pal and G.P. Mishra, Development, preparation and characterization of uranium molybdenum alloys for dispersion fuel application, J. Alloys and Compounds, Article In Press, On line available on sciencedirect.com.
17. A.G. Harding and M.B. Waldron, UKAEA (Harwell Report, AERE M/R-2673), part1 (1958).
18. J. Lehmann, Comptes Rend. Acad. Sci. Part 248 (1959) 2098.

❍❍

20

Improved Powder Metallurgy Technique for Fabrication of Mixed Uranium-Plutonium Carbide Fuel Pellets for Fast Breeder Test Reactor

Sudhir Mishra, P.S. Kutty, U. Basak and Arun Kumar Radiometallurgy Division Bhabha Atomic Research Centre, Mumbai, India*

Abstract

Mixed Uranium-Plutonium Carbide [($U_{0.3,}$ Pu0.7) C] in the form of cylindrical pellets is the fuel for Fast Breeder Test Reactor (FBTR) at Kalpakkam. For the last 25 years, these fuel pellets are manufactured at Radiometallurgy Division, Bhabha Atomic Research Centre, Mumbai by classical powder metallurgy route involving mixing of starting powder, vacuum carbothermic reduction, crushing of carbide clinker followed by milling in ball mill . Other processes involved are pre-compaction, granulation, final compaction and sintering. The total time for the completion of one batch comes to almost 90 hrs. Mixing and milling of oxide-graphite mixtures and milling of mixed carbide powders are found to be the major time consuming processes in order to get powder suitable for further processing. The entire manufacturing processes are carried out in a train of enclosed chambers of glove-boxes housing P/M process equipments. Recently, a new facility for fabrication of mixed carbide fuel pellets has been set up. This facility houses indigenously designed equipments like attritor, PID controlled higher capacity all metal furnaces, Programmable Logic controlled (PLC) double bore compaction press for final compaction besides other equipments. These equipments have been manufactured through constant guidance and discussion with local vendors. The use of these equipments has considerably reduced the process time. *Radiation exposure has also reduced significantly as the contact time of the personnel with equipment has minimized.*

20.1 INTRODUCTION

Three stage nuclear power programme pursued by India aims to utilize judiciously limited reserve of uranium and vast reserve of thorium. In accordance with this programme natural uranium is used in operating pressurized heavy water reactors (PHWRs) in first stage. In second stage, plutonium obtained from the reprocessing of PHWR spent fuel is planned to be used along with uranium as fuel and thorium as blanket material in fast breeder reactor. In fast reactor plutonium breeding takes place with simultaneous conversion of thorium into U^{233}. U^{233} shall be used as Th-U^{233} fuel in thermal reactors in third stage. Fast Breeder Test Reactor (FBTR) situated at Kalpakkam in India is a front runner for making the base of fast reactor technology in India. Commercial fast reactor named Proto type Fast Breeder Reactor (PFBR) presently under construction will be in operation in near future. Mixed Carbide fuel of composition $(U_{0.3},Pu_{0.7})C$ was indigenously developed and today this is the driver fuel for FBTR. This fuel in the form of cylindrical pellets is being fabricated in BARC. The reactor was made critical in year 1985. Recently a new fabrication facility has been commissioned in BARC with sate of art equipments developed and manufactured in India.

20.1.1 Fabrication of Mixed Uranium-Plutonium Carbide Fuel

Mixed $(U_{0.3},Pu0.7)$ C is pyrophoric material and is highly susceptible to oxidation and hydrolysis. The fabrication of carbide is therefore done in inert atmosphere inside glove box. The glove-box atmosphere is constantly monitored for oxygen and moisture and maintained $\leq$ 25 ppm each during fabrication process [1]. The carbide fuel is fabricated using classical powder metallurgy technique involving mixing of feed powder, carbothermic reduction of the mixed powder to obtain carbide clinker. The clinker is crushed and milled to obtain fine sinterable grade powder followed by precompaction, granulation and final compaction to get green pellets. The fabrication of carbide fuel for FBTR started in early 1980's. Since then till recent past the process flow sheet being followed is shown in the Fig. 20.1. The mixing of feed powders namely graphite, UO_2 and PuO_2 is done in planetary ball mill. It takes almost 12 hrs to get the powder properly homogenized. The mixed powder is subjected to process control check like determination of Pu/U ratio by XRF and C ananlysis by carbon determinator. The powder is tableted at 60 Mpa and loaded in to all metal bell type furnace of capacity 600 gms for carbothermic reduction under vacuum of the order of 10^{-3} mbar. The carbothermic reduction takes place as per following equation.

$$0.3UO_2 + 0.7PuO_2 + 3C = (U_{0.3}Pu_{0.7})\ C + 2CO \uparrow$$

The carbide clinker thus obtained is subjected to process control step for determination of C, O, Pu/U ratio N and M_2C_3 (M=U+Pu) in the product. After completion of reduction of one more set of 600 gms of mixed powder, the milling of total 960 gms of carbide clinker (20% loss in form of CO and CO_2) is done in planetary ball mill for 16hrs. This is followed by addition of zinc behenete and nepthalene as lubricant and binder respectively in the milled powder. Precompaction is done using hydraulic press of 10 tons and in a die of dia 12 mm. The normal pressure during precompaction is maintained at 150 Mpa. The powder is subjected to final compaction where pressing is done using single bore die to fabricate green pellets of dia 4.65 mm and height between 7 to 8 mm. The final compaction press is of 10 ton capacity and compaction pressure is maintained at 450 Mpa .The green pellets thus produced are stacked in tungsten trays and subjected to dewaxing followed by sintering at 1680°C in Ar+8% H_2 atmosphere.

20.2 NEW FABRICATION FACILITY

Recently a new fuel fabrication facility of mixed carbide has been set up. Significant modifications have been done in the process steps like introduction of attritor for mixing and milling, higher capacity all metal furnaces for carbothermic reduction (CR) and sintering, use of liquid binder and lubricant, multicavity die for final compaction etc. The changes in the flow sheet in Fig. 20.1 have been shown by dotted line boxes.

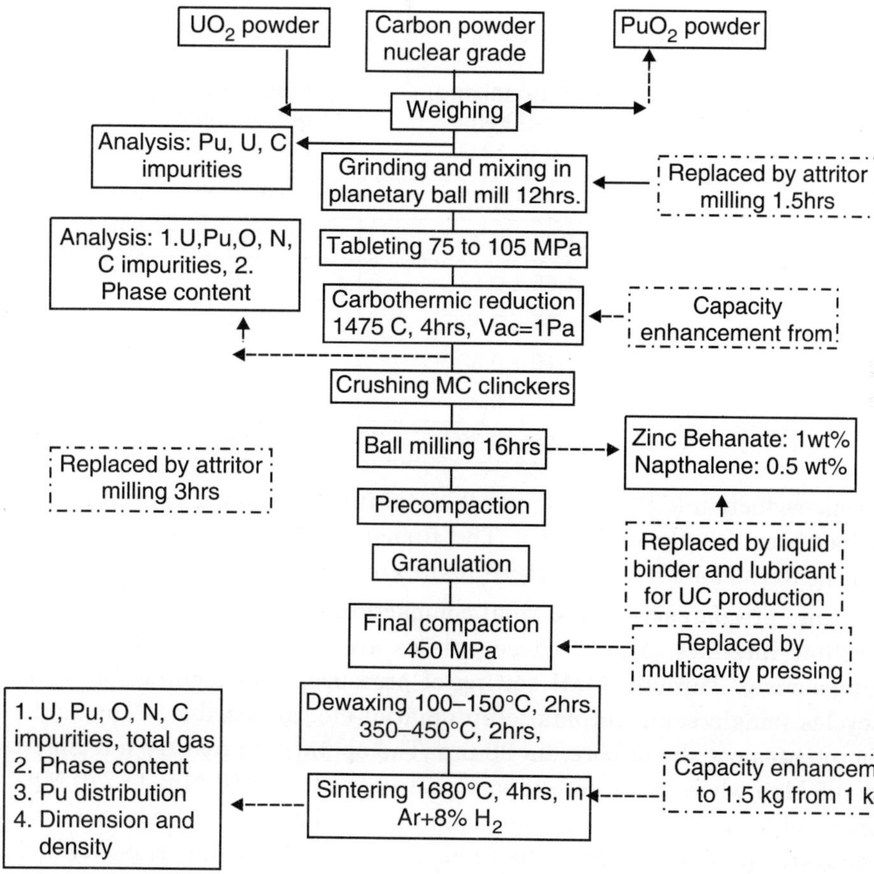

Fig. 20.1 Flow sheet for fabrication of (U, Pu) C pellets for FBTR

As per the modified process flow sheet mixing and milling of feed oxide and graphite powders and carbide powder is done in attritor machine. During mixing speed of attritor is kept low to avoid segregation of graphite powder. Graphite being light does not mix uniformly with UO_2 and PuO_2 when the speed is high. The design of the attritor has been done modular, so that it could be easily dismantled and assembled again inside the glove-box. The attritor is of 2 ltrs. capacity. At present charge of 1.2 kg is processed at a time. The separation of ball and powder takes place automatically during discharge by opening the bottom ball valve.

This is in contrast to planetary ball mill where intermittent stirring has to be done and separation of ball and charge has to be done manually. Because of high energy mixing effect in the attritor the time for mixing and milling as well as the contact time of personnel with equipment is reduced significantly.

Results of carbon analysis and Pu/U ratio of co-milled oxide–graphite powders in the attritor is shown in the Table 20.1. As observed from the table, the homgenised mixing of UO_2, PuO_2 and graphite powder occurs in one and half hour of mixing which otherwise used to take 12 hrs by ball milling.

Table 20.1: Analysis of oxide-graphite powder mixed in attritor

Batch No.	% C (after mixing) Control Value 11.60 to 12.00	Pu/U Ratio Control Value 2.40 to 2.50
1	11.70 ± 0.33	2.46
2	11.97 ± 0.32	2.50
3	11.65 ± 0.33	2.40
4	11.60 ± 0.33	2.44
5	11.70 ± 0.32	2.50
6	11.78 ± 0.33	2.48
7	11.98 ± 0.32	2.45

The carbothermic reduction (CR) is done in all metal furnace having capacity of 1.5 kg load. The furnace has been locally developed. The furnaces use tungsten charge carrier and tungsten mesh heater. The CR temperature maintained is 1490°C.

The tabeltting, precompaction and final compaction presses are operated through PLC and man machine interface (MMI). These presses are of 20 tons capacity and have been developed indigenously. Through MMI setting of pressure, data feeding for bore diameter, selection of cycles (single/semi automatic/automatic) etc. is possible. The tabletting and precompaction presses use die of bore dia 30 mm.The operating pressure for tabletting press is from 75 to 105 Mpa whereas for precompaction it varies from 150 Mpa to 270 Mpa. Because of larger tonnage of tabletting and precompaction press the die of 30 mm bore dia could be chosen to yield faster production rate. The final compaction press has capacity of 5 tons and uses double bore die of dia 4.60 mm. The operating pressure varies from 225 to 450 Mpa in order to meet specific mass to height ratio (linear mass).

The clinker produced after carbothermic reduction is milled in attritor for 3 hrs. The surface area of the powder increases with milling time as shown in the Fig. 20.2. It was also observed that with use of 10 mm tungsten ball as compared to 6 mm ball the specific surface area of the powder increases for the same time of milling.

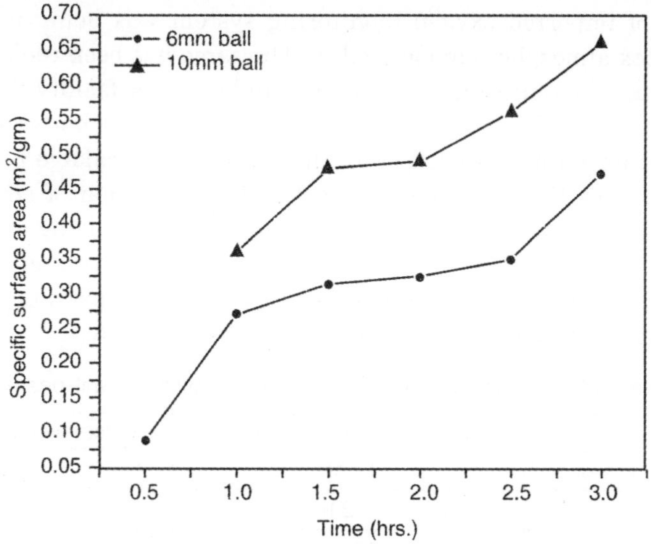

Fig. 20.2: Variation of specific surface area with ball size and time

In old facility we were using hydraulic press with single bore die .The carbide is non free flowing powder, the bore dia is also small so filling takes time. So press with two bore die in new facility has helped us to reduce the compaction time significantly reducing the contact time with machine and ultimately personnel dose.

The furnaces which we use for carbothermic reduction and sintering are all metal resistance type of furnace. The furnace operates at low voltage and high current. It has circular base with dome shaped hood sitting over it. The heater is mesh type cylindrical tungsten heater. The heater is kept on copper block which is connected to electrode for supplying power. In old furnace in our facility the copper block was connected to each other by a common water flow path for cooling purpose. This type of connection between two copper block leads to leakage of water in case of failure of "O" ring at the connecting point. In view of this problem in the new furnace each copper block has been given separate cooling line avoiding chance of water leakage. The dome of the furnace is heavy and it becomes difficult to lift it manually. The dome of the furnace was lifted manually inside the glove-box in the old facility. Motorised lifting arrangement has been incorporated in the new furnace which has made the system easy to operate. HEPA filter has been put in exhaust and gas line to avoid release of radioactive particulate. These filters have been installed inside the glove-box so that change of the filter can be done inside the glove-box easily restricting release of radioactivity. For isolation of furnace chamber with vacuum line/gas line bellow seal valves have been provided where ever necessary. This has given excellent performance. The oil cooled transformer has been replaced by compact air cooled transformer leading to better utilization of laboratory space. With the introduction of new transformer we are able to operate the heater at high current and this has also facilitated the use of perforated sheet type cylindrical molybdenum heater during sintering in A+H_2 atmosphere. These heaters are cheaper and have long life as compared to tungsten mesh type heater.

Analogue type of imported oxygen monitoring system was being used for continuous monitoring of glove-box atmosphere in old facility. The same has been replaced by indigenously developed digital type of monitoring system with audio alarm facility. The use of this type of monitor gives better watch over the purity of the box atmosphere as sudden change in the reading alerts us for immediate action for diagnosis of the leakage problem in the glove-box. Table 20.2 compares the total time taken for the production of sintered pellet as per old facility and new facility. It may be noted that time taken for mixing of powder and milling of clinkers reduces significantly to 1.5 hrs and 3 hrs from 12 hrs and 16 hrs respectively. Because of the use of two bore die, the time period of final compaction reduces by half. All together the total time of production for a batch reduces by ~30%.

Table 20.2: Comparison of time required for different unit operation

	Fabrication time in old facility	Fabrication time in new facility
Mixing of the feed powder	12 hrs	2 hrs
Tabletting	2 hr	1 hr
Carbothermic reduction	16 hrs	16 hrs
Milling of clinker	16 hrs	3 hr
Precompaction	1 hr	0.5 hrs
Final compaction	8 hrs	4 hrs
Dewaxing	20 hrs	20 hrs
Sintering	18 hrs	18 hrs
Total time	93 hrs	64.5 hrs

Table 20.3 shows result of important chemical analysis of the product made in different batches. As observed from the table the value of different chemical components is within the specified limit.

Table 20.3: Chemical analysis sintered pellets

Chemical Analysis for Different Batches	**1**	**2**	**3**	**4**	**5**	**6**	**7**	**FBTR Carbide Pellet Specification**
Pu+U (wt %)	94.50	94.90	94.60	94.76	94.92	94.77	94.81	≥ 94
Carbon (wt %)	4.82	4.92	4.90	4.94	5.01	4.95	4.85	≤ 5.03
Oxygen (ppm)	6000	5300	5800	5800	5850	4338	4454	≤ 7500
Nitrogen (ppm)	280	350	200	794	400	142	160	
M_2C_3 (wt%) (by XRD)	18	19	17	16	19	17	16	5 to 20%
Total impurities (ppm)	<500	<500	<500	<500	<500	<500	<500	≤ 3000

A comparative table of different equipments in new and old facility is as shown in the Table 20.4.

Table 20.4: Equipments in old facility and new facility

Equipments	Old Facility	New Facility
For Mixing	Planetary Ball Mill For 1.2 kg charge	Attritor Up to 2 kg charge
Tabletting	Hydraulic Press 10 ton capacity	PLC based Hydraulic press 20 ton capacity
Carbothermic Reduction	All metal Furnace having capacity 600 gm	All metal furnace of capacity 1.5 kg
Milling	Planetary Ball Mill for 1.2 kg charge	Attritor up to 2 kg charge
Precompaction	Mechanically operated Hydraulic press 10 ton capacity	PLC based Hydraulic Press 20 ton capacity
Final Compaction	Mechanically operated Hydraulic Press of 10 ton capacity	PLC based Hydraulic Press of 5 ton capacity
Sintering	All metal Furnace having capacity to load 1 kg pellet	All metal furnace having capacity to load 1.5 kg

20.3 FABRICATION OF UC INSULATION PELLETS FOR FBTR

UC Pellets are used as insulation pellets at both end of FBTR fuel pin. UC pellets have been fabricated using Liquid binder and lubricant. Polyethylene glycol (0.5 wt %) and oleic acid (0.3 wt %) was used as binder and lubricant respectively [2]. The use of liquid binder and lubricant obviates dewaxing step reducing the total production time by 50%.

Table 20.5: Sintered density variation with temperature and bore diameter

Batch No.	Bore Dia of Die (mm)	Surface Area m^2/gm	Green Density (%TD)	Sint. Temp(°C)	Sintered Density (%TD)	Sintered Dia of Pellet(mm)
1	4.65	0.78	70 at 20 TSI	1680 for 5 hrs	92– 93%	4.23–4.27
2	4.65	0.66	71-72 at 22 TSI	1660 for 4 hrs	92–93%	4.25–4.28
3	4.65	0.45	71-72 at 22 TSI	1625 for 4 hrs	92–93%	4.26–4.30
4	4.65	0.48	71-72 at 22 TSI	1600 for 4 hrs	91–92%	4.27–4.31
5	4.65	0.47	71-72 at 22 TSI	1575 for 4 hrs	87–89%	4.30–4.34
6	4.65	0.50	71-72 at 22 TSI	1550 for 4hrs	82–85%	4.39–4.46
7	4.50	0.52	71-72 at 22 TSI	1590 for 4 hrs	87–90%	4.15–4.18

Green pellets were fabricated using two bore die of dia 4.65 mm in double compaction mode. Aim was to achieve sintered density of 85–90%TD with dia ranging from 4.10 to 4.22 mm. As observed from the Table-20.5 for batch no.1 with sintering temperature of 1680°C, the sintered density of 92 to 93% TD was achieved .The average dia of sintered pellet was 4.25 mm. In subsequent batches the sintering temperature was gradually reduced to 1550°C. As observed from the Table-20.5, the density of the pellet reduced to 82–84%TD with sintered dia of 4.27 at 1550°C. As the sintered dia was not within the specified limit the die bore of dia 4.5 mm was chosen for further work. With change of tooling and maintaining a sintering temperature of 1590°C the required dia and sintered density could be achieved.

20.4 CONCLUSIONS

1. Mixing and milling time have reduced significantly with the use of attritor.
2. Use of liquid binder and lubricant obviates the dewaxing step in UC fabrication.
3. No manual separation of ball and powder is needed in mixing and milling with attritor so the radiation exposure of personnel reduces.
4. Total production time reduced significantly (by ~30%) with use of attritor and enhanced capacity equipments.

REFERENCES

1. C. Ganguly, P.V. Hegde, G.C. Jain, U. Basak, R.S. Mehrotra, S. Majumdar and P.R. Roy. "Development and Fabrication of 70%PuC-30% UC Fuel for the Fast Breeder Test reactor in India", *Nuclear Technology*, Vol.72, Jan.(1986).
2. Sudhir Mishra, S.N. Raul, A.K. Mishra, I.D. Godbole, Arun Kumar and H.S. Kamath, "Use of Poly-ethylene glycol and oleic Acid in Fabrication of Nuclear Fuel Pellets", *Trans. PMAI*, Vol. 29, pp. 87–89 (1999).

❍❍

21

Development and Characterization of Aluminium Based Hybrid Metal Matrix Composites for Light/Heavy Duty Applications

Anil K. Chaturvedi, K. Chandra, P.S. Misra, A.A.S. Ghazi, and M. Asif. Department of Metallurgical and Materials Engineering Indian Institute of Technology Roorkee, Roorkee, India

Abstract

The present study deals with the investigation of dry sliding wear behaviour of aluminium alloy based composites, reinforced with silicon carbide particles and solid lubricants such as graphite/Antimony Tri Sulphide (Sb_2S_3). The first one of the composites (binary) consists of Al with 20% silicon carbide particles (SiCp) only. The other composite has SiCp and solid lubricants: Graphite + Sb_2S_3 (hybrid composite) at solid state. Both composites are fabricated through P/M route using "Hot Powder Forging Technology". The density and hardness are measured by usual methods. The pin-on-disc dry wear tests to measure the tribological properties are conducted for one hour at different parameters namely load: 30, 50 and 80N and speed: 5, 7 and 9m/s. The tested samples are examined using scanning electron microscope (SEM) for the characterization of microstructure and tribolayer on worn surface of composites. The results reveal that wear rate of hybrid composite is lower than that of binary composite. The wear rate decreased with the increasing load and increased with increasing speed. The results of the proposed composites are compared with iron based metal matrix composites (FM01N, FM02) at corresponding values of test parameters. These iron based metal matrix composites are also fabricated by P/M route using 'Hot powder Forging Technology'. The comparative study reveals that the proposed composites have lower friction coefficient, less temperature rise and low noise level; however they have little higher wear rate. It is concluded that the hybrid composite has acceptable level of tribological characteristics with blacky and smooth worn surface.

21.1 INTRODUCTION

Discontinuous reinforced aluminum metal matrix composites (DRAMMCs) are a class of composite materials having desirable properties like low density, high specific stiffness, high specific strength, controlled coefficient of thermal expansion, increased fatigue resistance and superior dimensional stability at elevated temperatures etc., [1, 2]. These materials have emerged as the important class of advanced materials giving engineers the opportunity to tailor the material properties according to their needs. Essentially these materials differ from the conventional engineering materials from the viewpoint of homogeneity. In composites, controlled distribution of one or more reinforcement materials in continuous metal matrix phase is possible. Large majority of these composite materials are metallic materials reinforced with high strength, high modulus and brittle ceramic phases which can be either continuous in the form of fiber, discontinuous in the form of whisker, platelets or particulate reinforcements embedded in a ductile metallic matrix. In the last two decades, wear performances of DRMMCs reinforced with various reinforcements ranging from very soft materials like graphite, talc etc., to high hardened ceramic particulates like SiCp, Al_2O_3 etc., [3–6] have been reported to be superior to their respective unreinforced alloys. A good number of studies have been done on the Al/SiCp [4–6] and Al/Graphite [3, 7] individually. Traditionally, the external lubricant plays an important role in wear behaviour. Though the wear behaviour of Al/SiCp is good, the addition of natural lubricant like graphite enhances the self-lubricating capacity of the composites, which is essential in some of the applications where lubrication needs to be applied periodically, especially for wear parts which are difficult to access. Solid lubricant contained in the composite can be released automatically during the wear process and reduces the wear. The incorporation of graphite alone will give the desired solutions but various researchers have identified both experimentally and mathematically that the incorporation of graphite will reduce the mechanical properties of the composites [7], which is undesirable for the component used in structural and high-elevated temperature applications. The wear behaviour of Al-alloy reinforced with SiCp-Graphite particles has been documented. As a result, combination of high-hardened SiC particles and soft graphite were used in the present work to investigate the wear behaviour of the Al-alloy incorporated with SiCp and graphite particles for comparison with corresponding value of iron based metal matrix composites (Fe MMCs) [8].

21.2 MATERIALS

21.2.1 Disc

EN-32 steel disc has been used as brake rotor/brake drum with hardness of 60 HRC, machined from a commercial passenger car brake rotor. The inner diameter, the outer diameter and thickness are 180, 140 and 15 mm respectively. The surface is machined to an average roughness value of 1.5 μm (measured by P-6 stylus profile meter) which is same as the roughness value of sliding surface of the actual commercial brake rotor. The composition of the steel disc material is shown in Table 21.1.

Table 21.1: Composition of EN-32 steel disc

Constituent	C	Cr	Si	Mn	P	S	Mg	Fe
Percentage	0.45	0.25	0.27	0.65	0.04	0.04	0.25	Balance

21.2.2 Test Pins

The chemical formulation for preparation of aluminum metal matrix composites (Al-MMCs) is shown in Table 21.2. The Al-MMCs were manufactured through powder metallurgy route using "Hot Powder Forging Technology" developed by Lenin *at el.* [8]. Proposed composites were fabricated by using of Al-alloy as matrix reinforced with silicon carbide particles (size 34 μm) and solid lubricants (graphite and Sb_2S_3). The composition of powders (wt %) was mixed into two stages by mechanical alloying, followed by attrition and conical ball milling for two and eight hours respectively and mixture filled in designed capsule. The capsule was heated up to 450°C and soaked for one hour in tubular furnace. At this temperature, it was forged in channel die without application of inert gas or vacuum by friction screw press of 100 ton capacity. After forging, slabs were prepared by using of machining processes. They were reheated to 530°C in muffle furnace and repressed in hydraulic press at 20 bar. Homogenization process was done at 530°C for six hours.

Table 21.2: Chemical formulation of Al-MMCs (by wt.%)

Samples	Nature of Composite	SiC_p (34 μm)	Graphite (45 μm)	Sb_2S_3	Al alloy
A01	binary	20	--------	-------	80
A02	hybrid	10	6	3	81

21.3 EXPERIMENTAL DETAILS

21.3.1 Determination of Density (Archimedes Principle)

The basic method of determining the density of the specimen by measuring the ratio of mass and volume was used. In this paper the density of the specimens was estimated by Archimedes' principle is shown in Table 21.3.

Table 21.3: Physical and mechanical properties of Al-MMC

Test Pins	ρ	(BHN)
Al01N	2.80	89
Al02N	2.73	83

21.3.2 Brinell Hardness Test

Test material was indented with a 10 mm diameter hardened steel ball subjected to a load of 15.25 kg applied for 10 to 15 seconds. The diameter of the indentation left in the test material is measured with a low powered microscope. Brinell harness number is calculated by dividing the load applied by the surface area of the indentation. The diameter of the impression is the average of two readings at right angles and the use of a Brinell hardness number table can simplify the determination of Brinell hardness structures. Brinell hardness number of test pins is shown in Table 21.3.

21.3.3 Wear Testing of the Composites

A pin-on-disc test apparatus was used to investigate the dry sliding wear characteristics of Al-MMCs (Al01 and Al02) as per ASTM G99-95 standards. The wear specimens (30 × 7 × 7mm) were machined forged slabs, and polished for metallographic examination. The tests were conducted for one hour at different loads: 30, 50 and 80 N and sliding speeds: 5, 7 and 9 m/s at room temperature. The initial weight of the specimen was measured in a single pan electronic weighing machine with a least count of 0.0001g. During the test the pin was pressed against the counter part rotating against EN-32 steel disc by applying the load. A strain–gauged friction-detecting arm holds and loads the pin specimen vertically into a rotating hardened steel disc. The frictional traction experienced by the pin during sliding is measured continuously by PC-based data-logging system. After running through a fixed time period, the specimen were removed, cleaned with acetone, dried and weighed to determine the weight loss due to wear. The difference in the weight measured before and after the test gives the wear of the specimen. The wear rates were determined using the weight loss method. The frictional forces are recorded by mechanical transducer/load cell. Although the frictional forces vary with sliding time, an average value for the analysis was considered. Each test was conducted for a constant load and speed for one hour and similar tests were carried out for different speeds and loads to investigate the tribological behaviour. Scanning Electron micrographs (SEM) of the prepared Al-MMC samples are shown in Fig. 21.1. The distribution of particles throughout the matrix is found to be fairly uniform for both binary and hybrid samples.

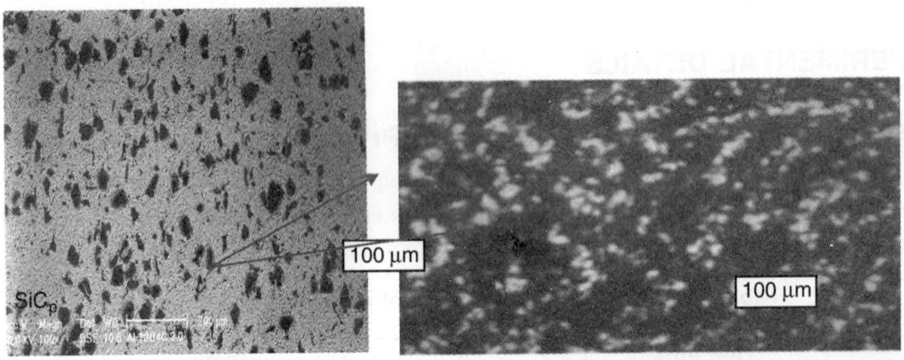

(a) (b)

Fig. 21.1: SEM micrographs of (a) Binary and (b) Hybrid MMC.

21.4 RESULTS

21.4.1 Effect of Load and Sliding Speed on Wear Characteristics

Figures 21.2(a) and (b) show the variation of wear rate (weight loss/min) with varying load and sliding speed for binary and hybrid composites. A drastic increment in wear rate of binary composite was observed with load increased from 50–80 N. The wear rate shows little variation with increasing sliding speed over the load range 30–50 N and for higher load range 50–80 N, slight reduction is observed with increased sliding speed from 07 to 09 m/s [Fig. 21.2 (a)]. For hybrid composite, it is almost stable and minimum at sliding speed 05 m/s and increased with increasing of speed from 05 to 09 m/s (30–50 N) and decreased for

50–80 N [Fig. 21.2(b)]. It has been concluded that wear resistance is improved with incorporation of SiC-particles with solid lubricants than that of binary Al-alloy.

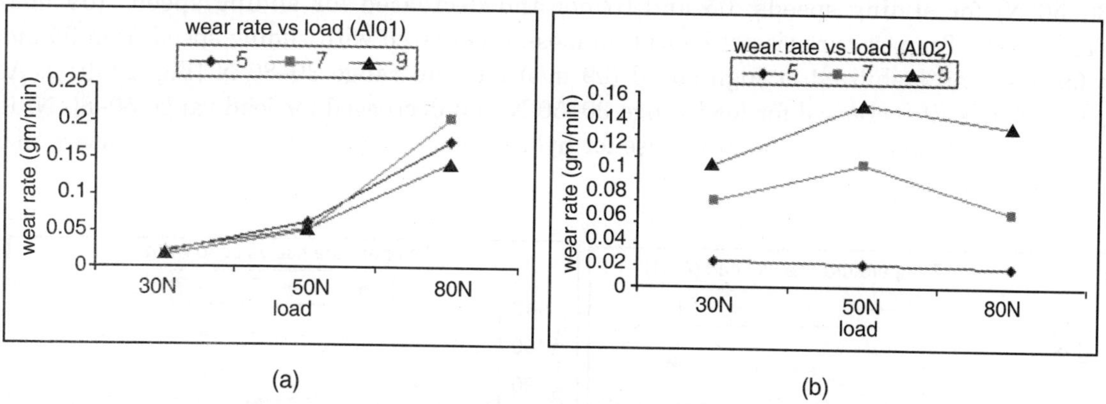

Fig. 21.2: Effect of load on wear rate of composites for different speeds 05, 07 and 09 m/s

21.4.2 Effect of Load and Sliding Speed on Co-efficient of Friction (COF)

Figures 21.3(a) and (b) show the variation of coefficient of friction with varying the load and sliding speed. The COF of binary composite decreased linearly with applied load (30–80 N) and speed (05–09 m/s) [Fig. 21.3 (a)]. It is observed that, the value of COF is high at low loads and tends to decease with increasing load, and nature of decline of COF with various sliding speeds is similar. For hybrid composite, the COF at low speed (05 m/s) is higher with reduced sensitivity at lower load range (30–50 N) [Fig. 21.3 (b)]. At speed (07 m/s), COF deceased with increased load from 30 to 50 N and becomes stable for higher applied load range (50–80 N). It is minimum at high speed (09 m/s) and decreased with increased applied load from 30–80 N. It has been concluded that coefficient of friction is stabilized with incorporation of solid lubricants for various sliding speeds and applied loads.

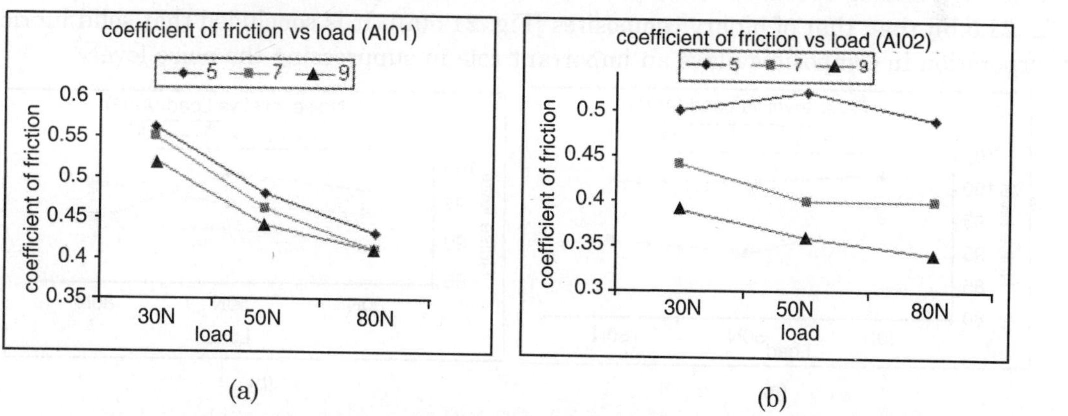

Fig. 21.3: Effect of load on coefficient of friction at different speeds: 05, 07 and 09 m/s.

21.4.3 Effect of Load and Sliding Speed on Temperature Rise at Mating Surface

The temperature rise of binary composite at mating surface is increased over lower load range (30–50 N) for sliding speeds; 05 and 07m/s and decreased for sliding speed (09 m/s) [Fig.21.4(a)]. Temperature rise of hybrid composite increased with sliding speed from 05 m/s to 09 m/s and stabilized at high speed (09 m/s) for load range 30–80 N [Fig. 21.4(b)]. At speed (05 m/s) it increased for load range 30–50 N and decreased for load range 50–80 N. It has been concluded that temperature rise for hybrid composite is more than temperature rise for binary composite.

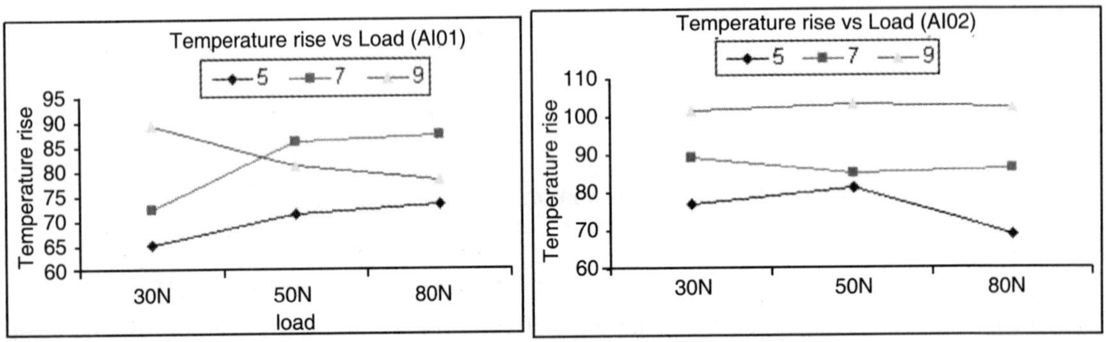

(a) (b)

Fig. 21.4: Effect of load on temperature rise for different speeds: 05, 07 and 09 m/s Effect of load on coefficient of friction at different speeds: 05, 07 and 09 m/s.

21.4.4 Effect of Load and Sliding Speed on Noise Level

From Figs. 21.5(a) and (b), it has been found that incorporation of complementary solid lubricants, the noise level is suppressed with reduced dampening and vibration. It is increased with load at sliding speeds; 07 and 09 m/s and decreased for speed; 05 m/s. It has been observed that at high sliding speed, the nose level is lower with slight variation for hybrid composites [Fig. 21.5(b)] than that of binary composites [Fig. 21.5(a)]. It is concluded that solid lubricant incorporation in composites plays an important role in suppressing the noise level.

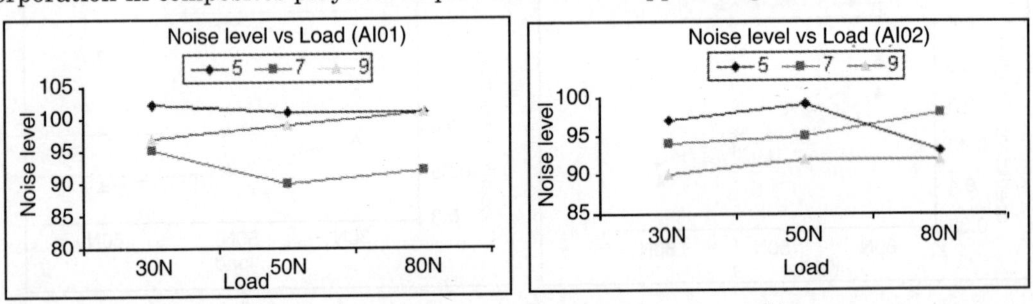

(a) (b)

Fig. 21.5: Effect of load on noise level of composites for different velocities: 05, 07 and 09 m/s

21.4.5 A Comparative Study of Al-Based Composites with Fe-Based Composites

Wear rate, coefficient of friction, temperature rise and noise level of aluminum based composites have been compared to corresponding values of iron based composites [8] at different loads for a sliding speed; 09 m/s. It has been found that the wear rate of aluminum based composites (Al01, Al02) is more than that of iron based composites (FM01N, FM02N). The variation of wear rate at high load for Al based composites is of opposite nature to the Fe- based composites [Fig. 21.6 (a)]. The coefficient of friction for Al-based composites is lower than that of coefficient of friction for Fe based composites at lower as well as higher loads, but nature of variation with load is similar [Fig. 21.6 (b)]. The temperature rise at mating surface is much lower in case of Al-based composites than iron based composites [Fig. 21.6 (c)]. The nature of temperature rise with load is not similar for both types of composites. For composites: Al02 and Fe01N, it is increased with load and decreased for composites: Al01 and Fe02N. The noise level is lower for Al-based hybrid composite than that of Fe based composite at higher load. It is slightly higher for Al-based binary composite at higher load applied [Fig. 21.6 (d)]. It has been concluded that the proposed Al-based composites with lower wear resistance have better tribological characteristics than Fe- based composites.

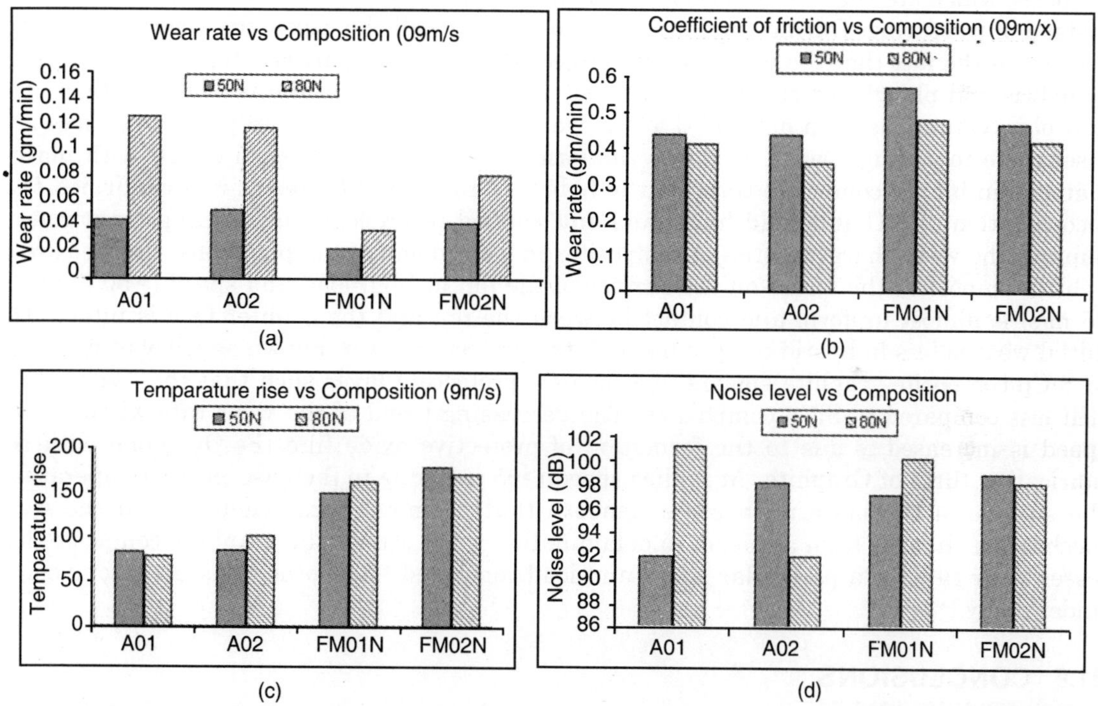

Fig. 21.6: Comparison of proposed Al based composites with corresponding values of Fe-based composites [8] at loads: 50 and 80 N for sliding speed: 09 m/s.

21.5 DISCUSSIONS

It was found from Fig. 21.1(a), (b) that wear rate of hybrid composite during run-in period is more than that of binary composite. The asperities of both the pin and counter face which are in contact with each other are subjected to relative motion under the influence of applied load. Initially both the surfaces are associated with a large number of sharp asperities and contact between the two surfaces takes place primarily at these points. In the present case also, the asperities on the pin are having a large number of reinforcements. Under the influence of applied load and speed, the asperities in each surface come in contact with each other and they are either plastically deformed or remain in elastic contact. As the asperities are very sharp in nature, the effective stress on these sharp points may be more than the elastic stress and then all these sharp asperities are plastically deformed at their contact points except the partially projected points of the reinforcement. The plastically deformed surface will fill the valley of the material both in pin and the counter face during the course of action and there is a possibility of fracture of a few asperities on both the surfaces leading to very fine debris. The asperities of the sliding pin surface come in contact with the steel disc surface and work hardening of the matrix material takes place under the applied load and speed [9]. The SiC particles, which are very strong in compression, are pushed back in to the soft Al alloy initially instead of fracturing. During the initial run-in period in all the cases, the wear rate is more because of the fact that the few of highly projected SiC broken particles from the pin will act as debris and plough the surface particularly the Al-particles. In the recent investigation a new observation has been marked that the wear rate during the run-in period is so less and insensitive to sliding speeds. It shows the formation of strong adhesion of SiC with matrix material in binary composite than that of hybrid composite. As a result it is confirmed that incorporation of SiC and solid lubricants and applied technology play an important role to improve the wear characteristics of composites in wear dominant applications. In the case of hybrid composites the stress on the surface of the pin at high load and speed (> 50 N and > 05 m/s) is almost uniform and contact between the pin and the counter face is intact. The initial wear is less in hybrid composites with the protecting layer, known as tribolayer, formed in SiCp composite-graphite smears and forms a protecting layer such that the wear rate is still less compared to SiCp composites. The decreasing trend of the wear rate when sliding speed is increased is due to the formation of protective oxide film (Fe_3O_4) along with the lubricating films of Graphite. At higher speed (>05 m/s) due to increase in the temperature, the surface of the material becomes smooth that promotes local yielding and the wear mechanism changes from abrasion to delamination wear. Hence the graphitic composite has lesser wear rate for a particular speed and load compared to all other classes of composites under study.

21.6 CONCLUSIONS

The following conclusions can be drawn from this work.

1. Incorporation of graphite particles in the aluminium matrix as a second reinforcement. decreases the wear rates of the composite compared to SiCp reinforced composite.
2. Coefficient of friction is stabilized with incorporation of solid lubricants in composition of composite at solid state for various sliding speeds and applied loads.

3. Temperature rise for hybrid composite is more than that of temperature rise for binary composite but it is significantly lower than iron based composites.
4. Incorporating solid ingredients in Aluminium powder play an important role to reduced the noise level.
5. Wear rate decreases as the sliding speed increases up to transition speed and load, due to work hardening of the surface, formation of Iron oxide, crushing of the SiC particles and smearing of Graphite.
6. Seizure occurred for wrought Aluminium based alloys, but no seizure occurred for Al/SiCp and Graphitic powder composites under the present study. Combination of abrasion, delamination and adhesive nature of wear was observed. Proposed aluminium based composites with lower wear resistance have better tribologocal characteristics than iron based composites.

REFERENCES

1. Ibrahim, I. A., Mohamed, F. A., Lavernia, E. J. Metal Matrix Composites. A Review *Journal of Material Science*, Vol. 26, pp. 1137–1157 (1991).
2. Sinclair, I., Gregson, P.J. Structural Performance of Discontinuous Metal Matrix Composites *Material Science and Technology*, Vol. 3, 1997: pp. 709–725 (1997).
3. Hsiao Yeh Chu, Jen Fin Lin. Experimental Analysis of the Tribological Behavior of Electroless Nickel-Coated Graphite Particles in Aluminum Matrix Composites under Reciprocating Motion *Wear* Vol. 239, pp. 126–142 (2000).
4. Lim, S.C., Gupta, M., Ren, L., Kwok, J.K.M. The Tribological Properties of Al-Cu/SiCp Metal Matrix Composites Fabricated using the Rheocasting Technique *Journal of Materials Processing Technology* Vol. (89–90), pp. 591–596 (1999).
5. Park, B.G., Crosky, A.G., Hellier, A.K. Material Characterization and Mechanical Properties of Al_2O_3-Al Metal Matrix Composites *Journal of Material Science* Vol. 36, pp. 2417–2426 (2001).
6. Grigoris Kiourtsidis, E., Stefanos Skolianos, M. Wear Behaviour of Artificially Aged AA2024/40μm SiCp Composites in Comparison with Conventionally Wear Resistance Ferrous Materials *Wear*, Vol. 253, pp. 946–956 (2002).
7. Mohan, S., Pathak, J.P., Gupta, R.C., Srivastava, S. Wear Behaviour of Graphitic Aluminum Composite Sliding under Dry Conditions *Wear*, Vol. 93, pp. 1245–1251 (2002).
8. Lenin S.D. Development of friction materials through powder metallurgy, Ph.D. Thesis, Indian Institute of Technology Roorkee (INDIA), 2008.
9. Korkut, M.H. Effect of Particulate Reinforcement on the Wear Behaviour of Aluminium Matrix Composites *Material Science and Technology*, Vol. 20, pp. 73–81 (2004).
10. Singh, M., Mondal, D.P., Modi, O.P., Jha, A.K. Two-Body Abrasive Wear Behaviour of Aluminum Alloy-Sillimanite Particle Reinforced Composite *Wear*, Vol. 253, pp. 557–368 (1993).

❍❍

22

Test Methodology to Evaluate the Wear Performance of PM Matrices used in Diamond Impregnated Tools for Cutting Hard Materials

L.G. Rosa, A. Coelho, P.M. Amaral, and J.C. Fernandes
Deptt. of Materials Engg., Inst. Superior Tecnico, Tech. University of Lisbon, Portugal

Abstract

To investigate the wear behaviour of PM matrices used in diamond tools and the possible relationship with their mechanical properties, an experimental study was carried out using circular discs to simulate the grinding process. Different metallic matrices were characterised, including the measurement of their mechanical properties. The wear behaviour of those matrices was evaluated during the processing of some common polycarbonate and soda-lime glass plates, and also in much harder materials, *e.g.*, granite tiles. The wear tests were run on special equipment available at the authors' premises, using discs with 125 mm in diameter, with a rotational speed of 2900 rpm and an infeed speed of 25 mm/s. The results showed a linear relationship of the wear with the normal force. The slopes of such linearity are related with the mechanical properties of the matrices, namely Young's modulus, toughness modulus, rupture stress and hardness. The wear rate (tool weight loss per run) of the matrices in granite is about 100 times higher than in glass; and in glass it is approximately 10 times higher than in polycarbonate.

22.1 INTRODUCTION

Abrasive machining processes are manufacturing techniques used to modify the shape and surface texture of manufactured parts by removing material from them [1]. The process employs very hard granular particles, like diamonds, and each diamond particle cuts away a small amount of material when forced into contact with the surface of the workpiece. These processes are capable of tighter tolerances and better surface finish than other manufacturing processes, but are usually expensive. In some industries, *e.g.*, ornamental stones, they represent the major cost of processing. Grinding is an example of abrasive process in which the abrasive particles are hold by a matrix to obtain a cutting element that is fixed, for example, to a wheel. The grinding wheel rotates at higher peripheral speed (typically in the range of 20 m/s to 45 m/s) compared to other machining processes such as milling and turning. The performance and life of a circular diamond tool are affected by many factors. Some of these major factors are: the characteristics of the workpiece material; the characteristics of the tool (*e.g.*, matrix, type of diamond, its size distribution and content); type of equipment and operator's skill. Understanding the tribological principles of these processes is crucial to reduce production costs, increase production rate, and improve surface quality in manufacturing. The wear of the matrix plays an important role, and therefore matrices have to be chosen properly. For example, if the matrix wears at high rate, the diamond is pulled out too soon and its capacity to cut is not completely used. On the other hand, if the matrix wears at a very low rate, there will be a continuous reduction of the gap distance between the surface of the matrix and the cutting edges of the diamond particles, leading to a significantly lower ability to cut.

The development of new metallic powders, applied as binding materials in the production of powder sintered tools containing diamond grit for cutting hard materials, originates the need to better characterise the behaviour of such matrix materials and their relation with the tool performance.

22.2 WEAR EQUIPMENT AND TEST METHODOLOGY

The wear tests were run using the "IST classification machine". Details of the present version of the test apparatus have been already presented elsewhere [2,3]. This equipment allows the use of circular tools (continuous or segmented) with a diameter between 100 and 400 mm; and wear tests can be conducted in several types of workpiece materials. The tools are coupled to an electric motor capable of giving them a constant rotational speed (between 500 and 2900 rpm). Another motor moves the tool (at a constant infeed speed, ranging from 5 to 25 mm/s) towards a platform to which the workpiece material is clamped. The equipment allows independent control of the following parameters: - rotational speed; - infeed speed; - depth of cut; - applied forces; - flow of coolant (liquid lubricant). The equipment also allows real-time acquisition of the following parameters: - vertical force (Fv); - horizontal force (Fh); - electrical energy (e) spent during the operation.

According to our methodology, the testing procedure establishes a certain number of cutting slots or runs. The objective is to determine the tool weight loss (.m) monitoring the resultant forces (generated by the contact between the tool and the workpiece material). The total number of runs depends on the characteristics of the workpiece material and type of tool. Figure 22.1 shows a view of the runs produced during a wear test conducted on a plate of soda-lime glass run/slot.

Fig. 22.1: Runs or slots produced during a wear test in soda-lime glass.

22.3 POWDER SINTERED MATRICES AND THEIR PROPERTIES

To investigate the wear behaviour of PM matrices used in diamond tools and the possible relationship with their mechanical properties, an experimental study was carried out using circular discs to simulate the grinding process. Seven metallic matrices were chosen in order to represent a wide range of metallic powders usually employed by the industry to manufacture the so-called "diamond tools" (see Fig. 22.2).

Fig. 22.2: Tested wheels; matrices without diamond (top), matrices with diamond (bottom).

The manufacturing process of the matrices comprises: 1) the formation of a green-body by uniaxial pressing of the mixed powders; 2) free sintering cycle during a pre-defined period of time (from 20 to 40 minutes) with plateau temperature in the range of 700 to 850°C; and 3) uniaxial pressure of the matrix immediately after the sintering cycle to achieve final densification, followed by cooling to room temperature. In order to evaluate the mechanical properties of the matrices, samples of each matrix were processed in a separate mould (with a parallelepiped cavity 80 × 10 × 10 mm). Some were machined to produce tensile test pieces like the one illustrated in Fig. 22.3.

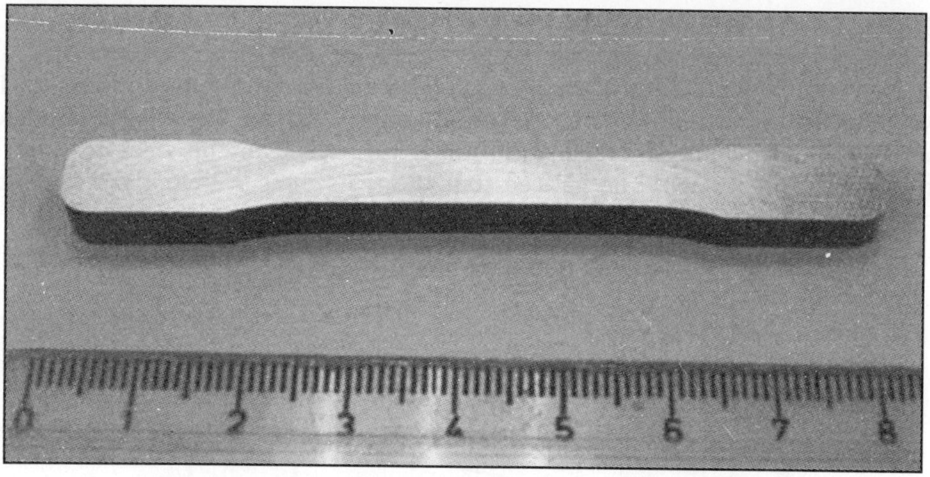

Fig. 22.3: Type of tensile test pieces of sintered matrix

The densities of the matrices (.) were evaluated by the Archimedes method. Vickers hardness (H) was determined using the hardness tester Mitutoyo AVK-C_2 with a load of 1 kg. Determination of the dynamic Young's modulus (E) was carried out using RFDA equipment in a flexural vibration mode [4]. Other mechanical properties: rupture stress (sr), rupture strain (er) and toughness modulus (Tm) were evaluated through tensile tests conducted in an Instron servo-hydraulic machine (Model 8502) using constant crosshead travelling speed of 0.5 mm/min.Chemical elements present in each matrix, density and mechanical properties of the seven matrices are presented in Table 22.1.

Table 22.1: Matrices used in this work; chemical elements, density and mechanical properties

Matrix Code Name	Chemical Elements	ρ [g/cm^3]	H [MPa]	E [GPa]	σ_r [MPa]	ε_r [%]	T_m [MJ/m^3]
LM89	Cu, Sn, Zn	8.62±162	1592±162	108.2±3.3	308±46	0.47±0.07	0.98±0.03
LM102	Cu, Sn, Co	8.49±0.05	2247±229	130.1±6.0	271±27	0.24±0.01	0.30±0.01
LM135	Cu, Sn, Co	8.30±0.07	1682±172	109.9±4.3	279±13	0.69±0.04	1.65±0.07
LM156	Cu, Sn, Co	8.22±0.08	1325±135	124.7±1.2	298±31	0.36±0.06	0.67±0.13
LM177	Cu, Sn	8.58±0.06	1500±136	101.7±1.2	192±35	0.22±0.06	0.28±0.07
LM189	Fe, Cu, Co	7.67±0.02	2677±273	215.2±4.1	222±25	0.22±0.02	0.46±0.08
LM193*	Fe,Cu, Co	8.09±0.02	2349±240	219.9±2.1	977±21	0.55±0.04	3.38±0.36

*this matrix is very similar to LM189, the difference comes from the powders suppliers.

22.4 RESULTS OF WEAR TESTS IN MATRICES WITHOUT DIAMOND

Experimental results obtained in plates of soda-lime glass are presented in Fig. 22.4 as plots of the wear (measured as tool weight loss per run, msc) versus normal force (Fn). Rotational speed of 2900 rpm and infeed speed of 25 mm/s were used in the wear tests of matrices without diamond, but it should be noticed that the tests in soda-lime glass were conducted in dry condition (*i.e.*, without liquid lubricant).

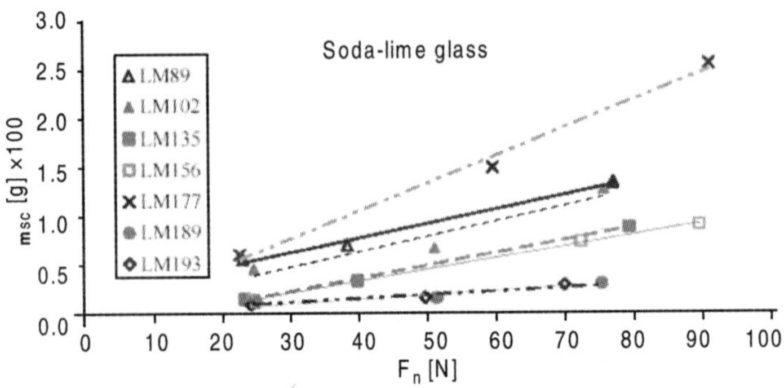

Fig. 22.4: Results of tool weight loss versus normal force in soda-lime glass

Recent investigation [5] has shown that the wear of the matrices can be related with their mechanical properties through the following empirical expression:

$$m_{sc} = \alpha \times \frac{F_n}{\sigma_r^{\,0.6} \cdot T_m^{\,0.1} \cdot H_V^{\,0.5} \cdot E} \qquad ...(1)$$

where msc represents the matrix weight loss per run and a is a constant of proportionality. The graphical representation of Equation 1 is presented in Fig. 22.5.

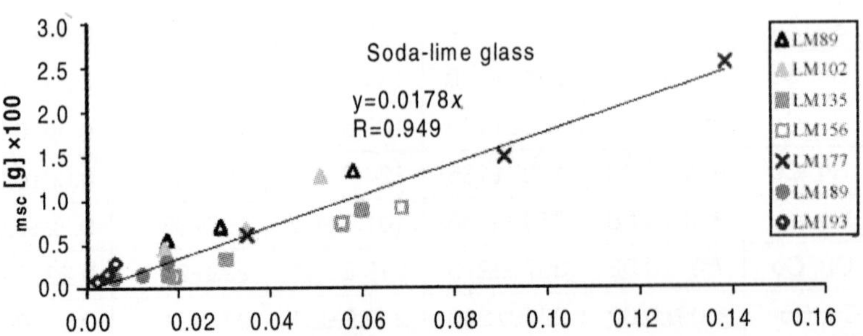

$$\frac{F_n}{\sigma_r^{\,0.6} \cdot T_m^{\,0.1} \cdot H^{0.5} \cdot E} \times 1000$$

Fig. 22.5: Graphical representation of equation 1; correlation for results in soda-lime glass

Equation 1 can also be applied to the results obtained from wear tests (of matrices without diamond) conducted in other types of materials, harder than soda-lime glass. Figure 22.6 shows the application of Eq. 1 to wear test results obtained in three kinds of commercial granite (portuguese "Azul Alpalhao", spanish "Porrino", and south-african "Impala").

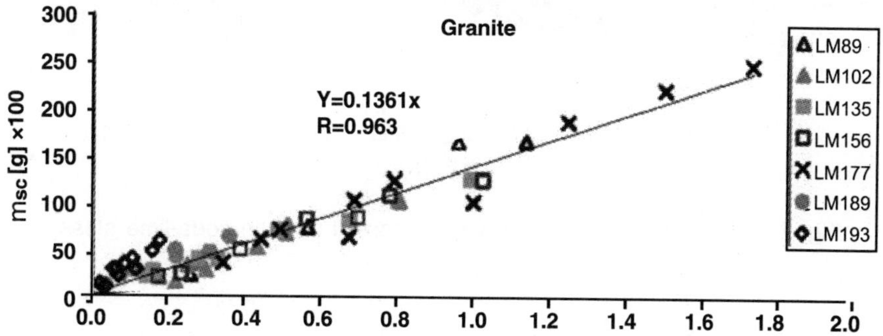

$$\frac{F_n}{\sigma_r^{0.6} \cdot T_m^{0.1} \cdot H^{0.5} \cdot E} \times 1000$$

Fig. 22.6: Graphical representation of equation 1; correlation for results in three kinds of granite

22.5 TESTS USING MATRICES WITH DIAMOND

Results obtained (also at 2900 rpm; infeed speed = 25 mm/s) using matrices with diamond are presented in Figs. 22.7, 22.8 and 22.9, respectively for tests conducted on extruded plates of Makrolon® (a high-tech polycarbonate from Bayer), on soda-lime glass plates, and on "Porrino" granite tiles. Qw represents the removal rate (in 3 mm per second) of workpiece material (*i.e.*, polycarbonate, glass, or granite).

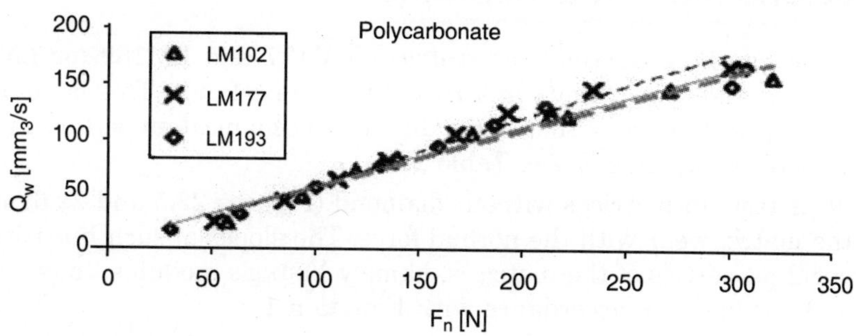

Fig. 22.7: Material removal rate versus normal force in Makrolon® polycarbonate

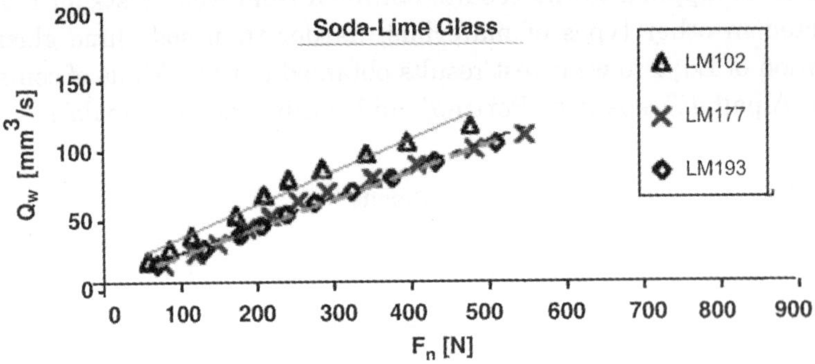

Fig. 22.8: Material removal rate versus normal force in soda-lime glass

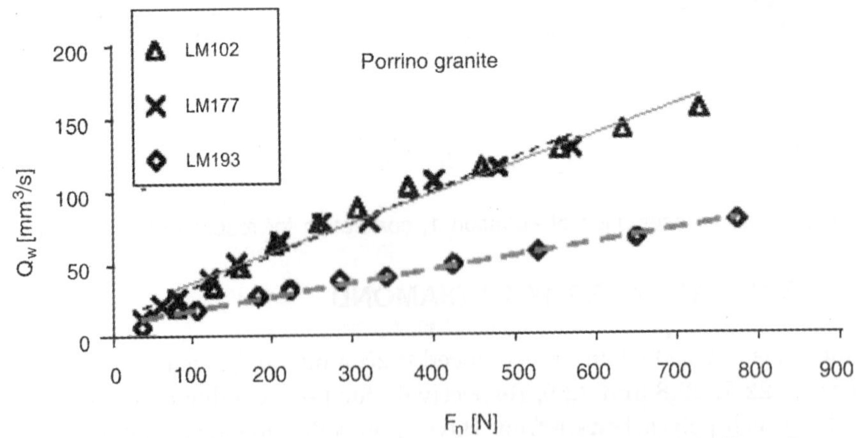

Fig. 22.9: Material removal rate versus normal force in "Porrino" granite

22.6 CONCLUSIONS AND FINAL REMARKS

From Figure 22.4 it can be observed that matrices LM177 and LM189 (or LM193) define, respectively, the lower and upper limits in terms of wear resistance. This is not surprising in view that those two types of matrices show quite different mechanical properties (mainly, Vickers hardness H and Young's modulus E, see Table 22.1).

Results of wear tests in matrices without diamond (Figures 22.5 and 22.6) show a linear relationship of the matrix wear with the normal force. The slopes of such linearity are related with the mechanical properties of the matrices, namely Young's modulus, toughness modulus, rupture stress and hardness, in accordance with Equation 1.

This paper is only a part of a much larger research study in which other wear parameters were determined. However, when the tests are conducted on workpieces of polycarbonate, the wear of PM matrices can not be adequately characterized because this type of polymer is much "softer" than the matrices. In practice, the wear rate (weight loss per run) of these PM matrices is approximately 10 times lower in polycarbonate than in soda-lime glass.

Apart from the results of wear in matrices without diamond, Figs. 22.7, 22.8 and 22.9 demonstrate the capability of the commercial tools (PM matrices with diamond) to adequately process much harder materials, like granite, at comparable material removal rates. For example, a diamond tool with matrix LM177 is capable of a material removal rate Q_w = 140 mm^3/s in both "hard" granite and "soft" polycarbonate. The only difference are the generated forces.

The experience gained with this line of research shows that the methodology for the evaluation of the matrix wear, as well as diamond tool performance, is quite satisfactory for a wide range of applications. The procedures that we use guarantee the conditions of repeatability of results.

REFERENCES

1. I.D. Marinescu, W.B. Rowe, B. Dimitrov, I. Inasaki, Tribology of Abrasive Machining Processes, William Andrew Inc., N.Y., pp. 1–3 (2004).
2. P.M. Amaral, J.C. Fernandes, L.G. Rosa, "Wear mechanisms in materials with granitic textures – applicability of a lateral crack system model", Wear, doi:10.1016/ j.wear.2008.08.018. Available online since 5 November, 2008.
3. J.C. Fernandes, S. Pinto, R. Rodrigues, P.M. Amaral, L.G. Rosa, "Development of a new testing method for assessing the wear behaviour of circular cutting discs", Materials Science Forum, Vols. 587-588, pp. 966–970 (2008).
4. L.G. Rosa, P.M. Amaral, J.C. Fernandes, "Experimental determination of Young's modulus in PM metal matrices used in diamond impregnated tools for cutting hard materials", Powder Metallurgy, Vol. 51, pp. 46–21 (2008).
5. A. Coelho, Avaliação do desgaste em matrizes metálicas usadas em ferramentas diamantadas e sua relação com as propriedades mecânicas, MSc thesis, Instituto Superior Tecnico, Technical University of Lisbon (2008).

○○

23

Machining/Enhancing Solutions for PM Materials

Ulf Engström and Olof Andersson, Höganäs AB, Sweden

23.1 INTRODUCTION

Powder metallurgy (PM) is known for producing complex parts to very close tolerances without the necessity of machining operations. However, machinability is still important for many applications. Many components require machining to reach final shape due to particular geometries (such as holes perpendicular to the pressing direction, bevels, slots and threads), and also due to demands for even finer tolerances or surface-finish.

In this presentation an overview of the development that has taken place during the last decades with machinability of sintered steels is described in the form of guidelines for turning and drilling of some common PM materials.

The first part gives a short introduction to PM, process, needs of PM materials, typical machining operations and factors influencing machinability and typical alloy systems and their properties. The second and third parts deal with the influence of additives and processing respectively. Finally some conclusions and recommendations for machining of PM steels will be given.

23.2 DESCRIPTION OF THE PM PROCESS

The production of PM parts is schematically outlined in Fig. 23.1 above. It consists of three main processing steps: mixing, pressing, sintering. Initially, iron powder and various alloying powders are blended together with a lubricant. The powder mix is then compacted with a pressure in the range from 600 up to 1000 MPa in a press to produce components of a desired shape. The compacted components are then sintered in a protective atmosphere in a furnace for 20–40 minutes at 1100°C to 1300°C.

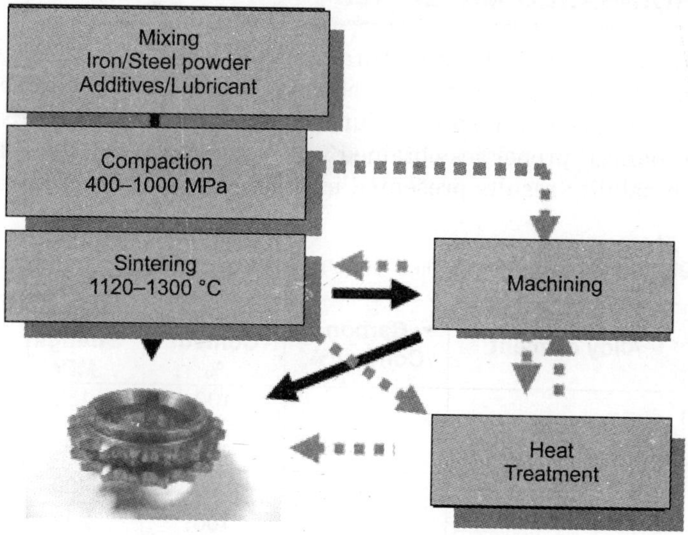

Fig. 23.1: Description of PM process

23.3 MACHINING OPERATIONS ON PM STEELS

As a metal-working technology, PM has the advantage compared to other processes, that it can produce complex parts of high quality with close tolerances, in an economical way. This is due to low energy consumption, high material utilization and low capital cost for the technology. Additional advantages are its high flexibility and particular mechanical properties, related to microstructure.

In order to improve the dimensional accuracy of the components, a second pressing operation (or coining) is often performed. Finished components can be hardened, machined, plated, et cetera, in the same way as ordinary steel components. Iron and steel powders for the manufacture of sintered structural components including sintered porous bushings are produced in many countries. World consumption of such powders increased rapidly during the past three decades and were about 1 000 000 metric tons year 2008.

Investigations of the PM market reveals that about 60% of all components need some kind of machining operation. Machining operations can account for up to 20% of the total production cost of a component. Drilling, tapping, and discontinuous turning, are considered the most difficult. Turning is by far the most frequently occurring operation, followed by grinding and drilling.

23.4 FACTORS THAT AFFECT MACHINABILITY

The machinability of a PM component is dependent on both work-piece and tool material as well as cutting conditions. Chemical composition, porosity, free machining additives, and production process parameters such as compaction and sintering methods, also all influence machinability.

23.5 SINTERED IRON-BASED MATERIALS

Alloy compositions of sintered steels for structural parts have to be carefully selected to meet the demand of the application in the most cost effective way not only with respect to desired mechanical properties but also with respect to dimensional stability during processing. In Table 23.1 below typical mechanical properties obtained with some commonly used alloying systems are shown. Most machinability results presented in this presentation are based on these alloy systems.

Table 23.1: Mechanical properties of commonly used PM alloying systems

P/M Group	Alloy element	Carbon Content %	Ferrite Content %	Tensile Strength MPa	Hardness HV10
ASC 100.29 (Ironbased)		0	100	180	60
	0.45P	0	100	400	100
	2Cu	0	100	280	70
	2Cu	0.25	85	400	120
	2Cu	0.5	30	490	130
	2Cu	0.8	1	600	150
Distaloy AE	4N1,1.5 Cu,0.5 Mo	0	90	360	100
	4N1,1.5 Cu,0.5 Mo	0.25	35	550	160
	4N1,1.5 Cu,0.5 Mo	0.5	8	620	200
	4N1,1.5 Cu,0.5 Mo	0.8	2	590	230
Astaloy Mo	1.5 Mo	0.5		500	160
Astaloy A+2% Cu	1.5 Mo, 4Ni, 2 Cu	0.5	<1	900	270
Astaloy CrM	0.5 Mo, 3Cr	0.5	<1	1130	480
(Pre-alloyed Mo, Cr, Ni)					

23.6 MICROSTRUCTURES

Microstructures of sintered alloyed steels, produced from powder mixes, are typically much more heterogeneous than those of conventional alloyed steels. While carbon diffuses rapidly in iron lattice and reaches equilibrium during sintering, other alloying elements like copper, nickel and molybdenum diffuse slower and reaches equilibrium after longer sintering times. Hence, when produced under commercially acceptable sintering conditions, these materials will always exhibit a certain degree of heterogeneity.

Plain iron has a pure ferritic microstructure. Pure ferritic material is difficult to machine due to smearing on tools. If carbon is added the microstructure will be a mix of ferrite and pearlite. This microstructure has the best performance regarding machinability. Formation of pearlite is a consequence of carbon addition and varies with the amount added. A typical microstructure produced with an iron-carbon alloy with 0.2%C can be seen in the first figure on the next slide.

The addition of graphite to iron-copper alloys has the very useful effect of increased strength. The next figure shows the microstructure of an iron-copper-carbon alloy with 2%Cu and 0.6%C. The microstructure is practically identical with those of corresponding conventional plain carbon steels.

The Fe-Cu-Ni-Mo-C system covers a wide range of iron alloy powders available for PM applications. The microstructures of two of the most common of these grades, Distaloy AE and Distaloy HP-1, are shown in the other two microstructures. As can be seen these microstructures are heterogeneous contaning phases such as ferrit, pearlite, bainite, austenite and martensite in different amounts.

The completely pre-alloyed materials Astaloy Mo and Astaloy CrM will obtain a homogeneous microstructure as the alloying elements are well distributed within each particle from the start. The microstructure generated is depending on the carbon content and the cooling rate.

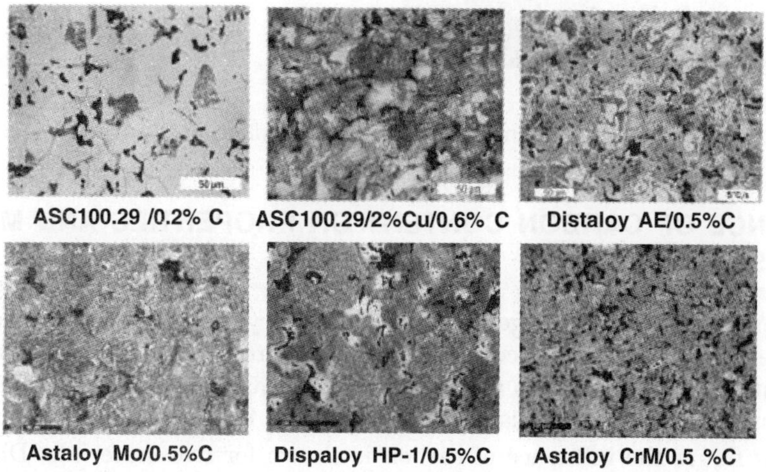

Fig. 23.2: Typical microstructures of common PM alloys (Magnification 300x)

23.7 MACHINABILITY TESTING

Machinability is not a universally defined property. Generally, it is the ability of a workpiece of material to be machined, meaning how easy or demanding it is to shape a workpiece with a cutting tool. Machinability can be divided into three main aspects [1, 2, 3, 4]:

1. Tool life.
2. Surface finish.
3. Power required to cut.

In order to state cutting parameters for machining of PM materials these three aspects must be investigated. In the following developments that been made with machining of PM steels for the two most frequent machining operations used in PM, turning and drilling, will be described. The test procedures are outlined in Fig. 23.3.

Turning tests were performed in a SMT turning lathe. Geometry of the working piece: inner diameter 35 mm, outer diameter 64 mm, height 62 mm. To illustrate intermittent turning a synchronising hub was used. All turning was performed as a facing operation.

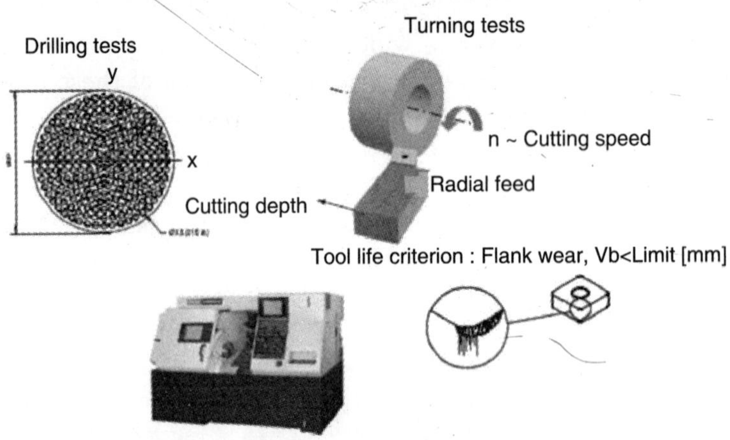

Fig. 23.3: Machining test procedures, drilling and turning

23.8 INFLUENCE OF CARBON CONTENT ON PROPERTIES AND MACHINING PROCESSES

Interstitial elements like carbon generally diffuse very rapidly in the iron lattice. As a consequence pearlite forms in the central part of the iron particle. This structure, from a machining point of view, is "easy" to work with. Depending on the type of tool and machining operation, an optimum amount of carbon can be stated for particular materials.

In the next figure the influence of carbon is shown for two materials, Distaloy AE and ASC100.29 +2% Cu. In turning ASC100.29+2% Cu without carbon addition, the influence from the proportion of ferrite is obvious. But note that as the amount of added carbon is increased beyond 0.25%, tool life decreases towards that occurring without added carbon.

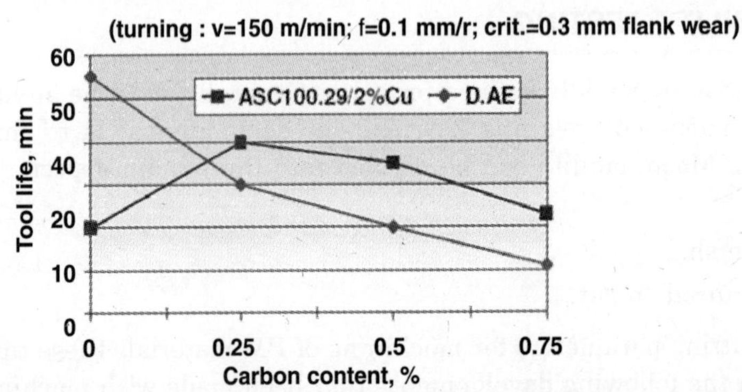

Fig. 23.4: Influence on tool life in a cutting operation of carbon content in Distaloy AE and ASC100.29 +2% Cu. Tool: GC1015 MF

Regarding the effect on Distaloy AE, the same positive influence from carbon addition could be expected due to the fact that the microstructure contains 95% ferrite, but this is not the case. The hardness of Distaloy AE is 100 (HV10) and the deviation from the hardness relationship shown previously is apparent. Amount of martensite increases with carbon addition. This particular phase can explain the decrease in machinability for the Distaloy grade when any amount of carbon is added.

23.9 CLASSIFICATION OF PM MATERIALS REGARDING MACHINABILITY

In conventional steel hardness is one of the main properties to define machinability. Low values are normally favourable. Exceptions are very ductile materials where problems arise from the formation of built up edges in the form of poor surface texture, burr formations and short tool life.

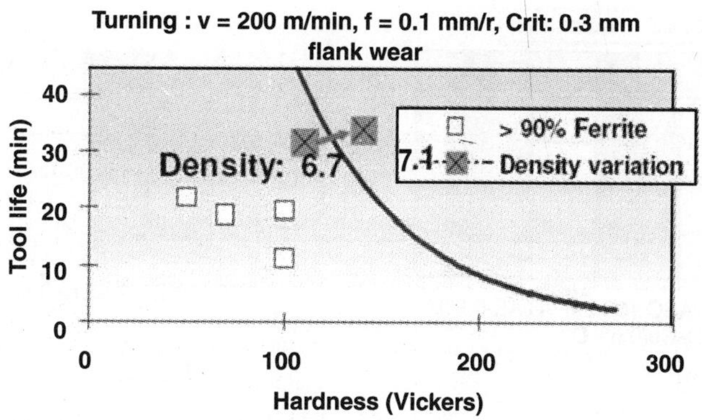

Fig. 23.5: Influence of hardness on tool life during turning (facing with machined surface: velocity 200 m/min, feed rate 0.1 mm/r, tool life criteria 0.3 mm flank wear)

Tool life was evaluated for the materials shown in Table 23.1 at a cutting speed of 200 m/min with a feed rate of 0.1 mm/r and criterion 0.3 mm flank wear [5] . Density of the materials was 7.0 g/cm^3. The tool used was CNMG120408, PVD-TiN coated hard metal. The results are shown in Fig. 23.5. If the amount of ferrite presence in the microstructure is taken into account PM materials can initially be classified by the macro hardness. By increasing the density from 6.7 to 7.1 g/cm^3 an improvement of the machinability can be achieved.

23.10 ADDITIVES

A large number of additives are reported to improve the machinability of P/M materials [6]. These additives must fulfil two main requirements:

- Improve machinability.
- Mechanical properties and dimensional change at the same levels as for material without additive.
- Easy to add without any special treatment

- Safe and environmental friendly
- Cost effective

Two additives, MnS and MnX, that have been developed to meet the above requests will be described in this paper.

23.11 INFLUENCE OF MnS AND MnX ADDITIONS

Evaluation of the performance of three different materials revealed a large influence from MnS with plain iron. The effect decreased with the amount of alloying elements as can be seen from Fig. 23.6(a). Lubrication on the cutting edge, improved chip breaking, and a reduction in cutting force are believed to be likely reasons for the improvement seen with addition of MnS.

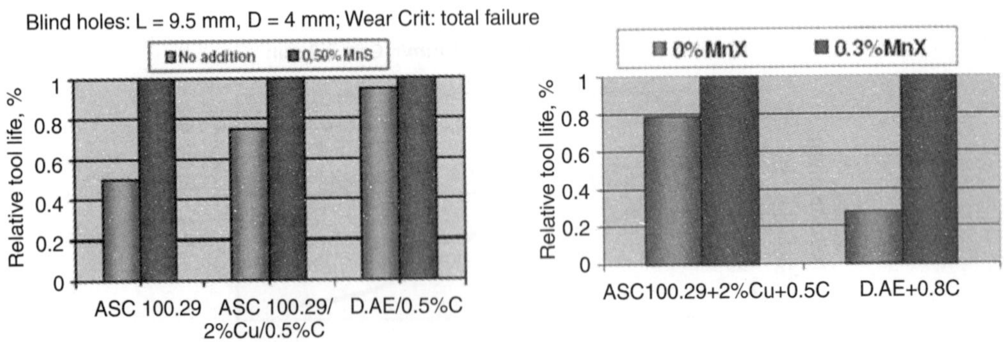

(a) (b)

Fig. 23.6: Influence of MnS (Fig. 23.6a) and MnX, (Fig. 23.6b) on tool life in a drilling operation of ASC100.29, ASC100.29 +2% Cu and Distaloy AE. Tool: GC1015 MF; velocity: 150 m/min; feed rate: 0.1 mm/r

As can be seen from Fig. 23.6(b) a large influence was obtained with MnX for both materials and especially for the Distaloy AE material.

23.12 CONTINUOUS AND INTERMITTENT CUTTING

The effect of MnS has been investigated regarding performance both in intermittent and continuous cutting of Distaloy AE/0, 5%C. The result of the continuous cutting test is shown in Fig. 23.7 (a) below. As can be seen increased cutting speeds resulted in an increased tool life by the MnS addition.

For the intermittent cutting test, turning of a synchronizing hub was used. It was found that MnS added to a Distaloy AE+0.5% C mix gave substantially longer tool life irrespective of changes in cutting speed as can be seen from Fig. 23.7 (b). In summary regarding the two tests, it can be seen that increased cutting speeds result in increased tool life and is further improved with MnS.

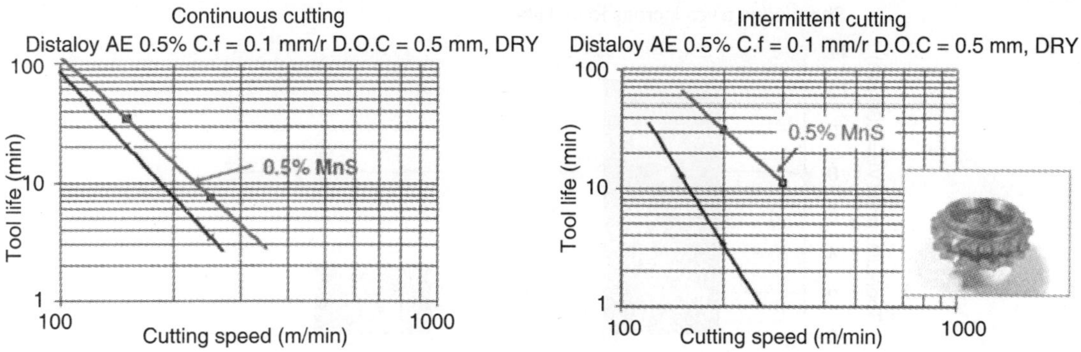

(a) (b)

Fig. 23.7: Influence on tool life shown as vT-curves of MnS addition to Distaloy AE/0,5%C, turned continuously with GC1015MF tool and intermittently with a GC3025 tool

Influence of MnS and MnX additions on turning tests of bainitic astaloy CrM

The influence of the additive MnS and MnX to Astaloy CrM with a bainitic microstructure is dramatically shown in Fig. 23.8. The machinability tests show that with tool insert CNMG120408-WF and cutting data (Vc = 150 m/min, f =0.1 mm/r, ap = 0.5 mm) the presence of BUE and smeared material on clearance flank was non-existent for the materials containing the additives, even when the machining took place without cutting fluid, *i.e.* dry.

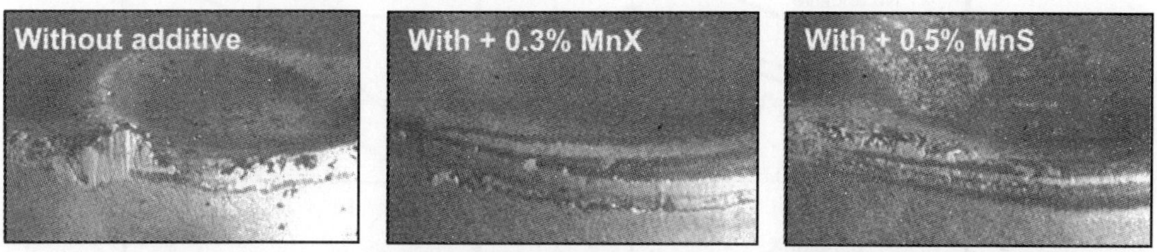

Fig. 23.8: Influence of MnS and MnX on tool life on dry machining of bainitic Astaloy CrM.

23.13 EFFECT ON MACHINING PARAMETERS

Both the tool and the cutting conditions have a large effect on the tool life. Some of the important parameters are listed below and some will be described in the following.

Tool: Material, Surface treatment, Breaker, Breaker shape, Cutting edge shape, Escape angle
Cutting Conditions: Feed rate-Cutting depth-Cutting angle-Cutting speed–Coolant

Influence of micro geometry for the insert on tool life and cutting forces

For PM materials sharp tools have for a long time been known as a condition for satisfactory machining. Figure 23.9 shows that the tool life is substantially increased for Distaloy AE by using a sharp tool.

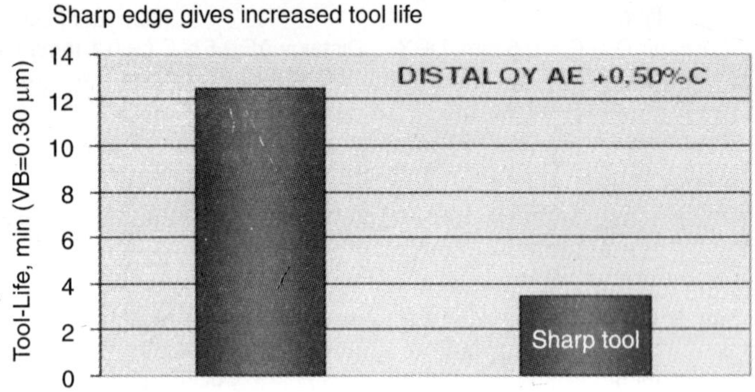

Fig. 23.9: Influence of tool sharpness on tool life

23.14 INFLUENCE OF MnS AND MnX ADDITIONS ON TORQUE AND FEED FORCE

Investigation of the cutting force and torque when drilling, as influenced by MnS addition, reveals a decrease in both cutting force and torque. For high strength materials, the effect of MnX is greater.The effect of MnX is in the same range or better compared to 0.5% MnS addition to Distaloy AE/0.5 C in regard to feed force (thrust) and torque as can be seen from the two diagrams in Fig. 23.10 below.

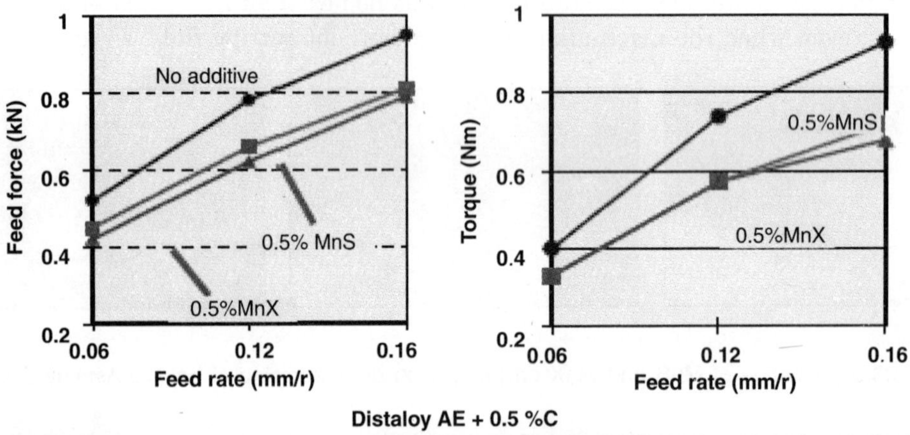

Fig. 23.10: Influence of MnS and MnX onfeed force and torque in a drilling operation of distaloy AE

23.15 INFLUENCE OF TOOL MATERIAL, CBN, ON SINTER HARDENED ASTALOY CrM

In Fig. 23.11 the results from a hard turning machining operation with a sintered hardened Astaloy CrM material with martensitic microstructure is shown. In this operation hard turning utilising a CBN insert was carried out. As can be seen from the left diagram the flank wear is very low up to about 7000 meters cutting distance. After that it start to wear more severe. On the other hand, which could be seen from the right diagram the surface finish is still very good after a cutting distance of 9600 meters.

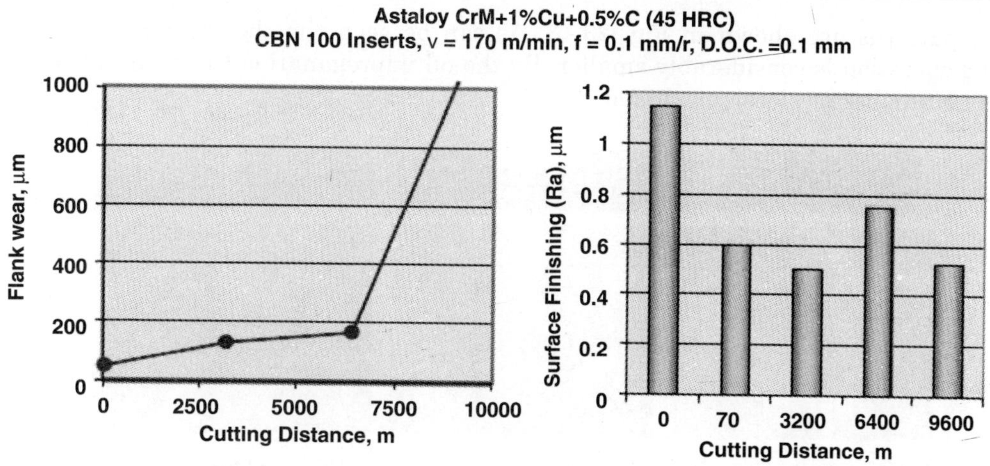

Fig. 23.11: Influence on tool wear by a hard turning machining operation with a sintered hardened Astaloy CrM material (martensitic microstructure)

23.16 INFLUENCE OF CUTTING MODE AND OIL IMPREGNATION

Typical for PM steels is the presence of porosity. By impregnation this porosity with oil the machinability can be substantially improved. Oil impregnation also increases the possibility of obtaining the greatest benefit from wear resistant, coated CVD tools.

Use of coolant during cutting is common with conventional steel. The situation is different with P/M materials since those have pores in the micro structure. Porosity combined with the presence of water can be detrimental to mechanical properties due to the resulting oxidation. Whenever possible the use of coolant and cutting fluid should be avoided in turning operations of PM materials. From a tool life point of view it is usually better to machine dry. This is particularily true with high performance grades such as coated CBN and CVD tools.

The effect of cutting mode on tool wear in cutting ASC 100.29/2Cu/0.5C and Distaloy AE/0.5C materials dry, wet and dry combined with oil impregnation is shown in Fig. 23.12. As can be seen the flank wear is markedly smaller with continued dry cutting compared to wet cutting. By combining dry with oil impregnation the improvement is even larger.

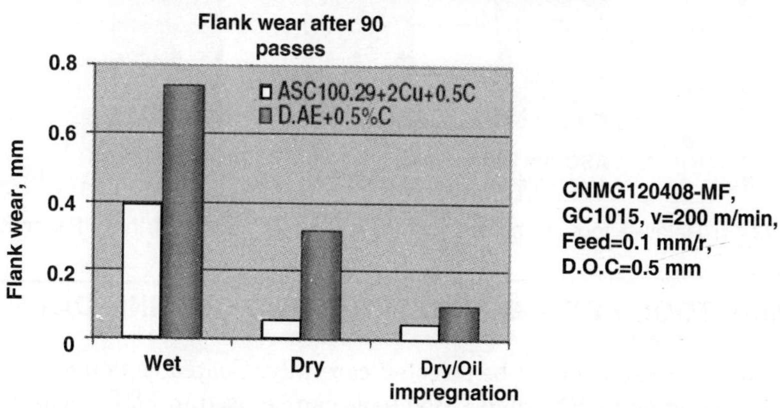

Fig. 23.12: Influence on tool wear at three different cutting conditions: dry, wet and dry combined with oil impregnation

The wear patterns are shown in Fig. 23.13. As can be seen the the wear pattern for dry machining operation is considerably smaller. By the oil impreganation the wear is very small at the same number of pieces machined.

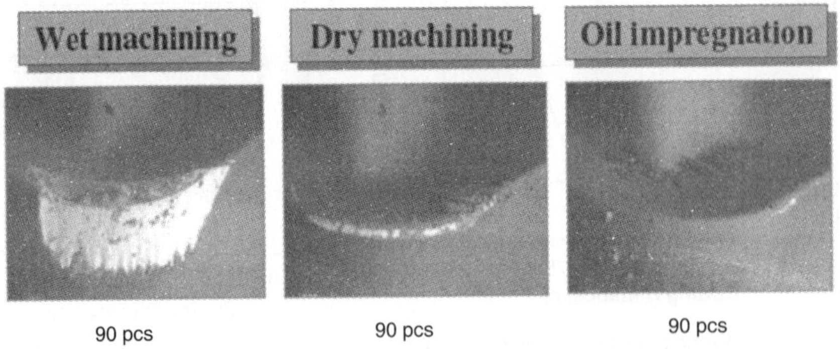

Fig. 23.13: Wear patterns obtained for different cutting modes

The obvious deterioration in tool performance with cutting fluid can be traced back to the severe thermo-mechanical load cycle associated with the test mode.

23.17 DRY VS WET DRILLING OF DIFFERENT MATERIALS

The application of cutting fluids or liquid coolants generally improves drilling performance as can be seen in Fig. 23.14. If the additional operation to prevent oxidation can be performed, taking the increased cost into account, the use of cutting fluids has a beneficial effect for drilling.

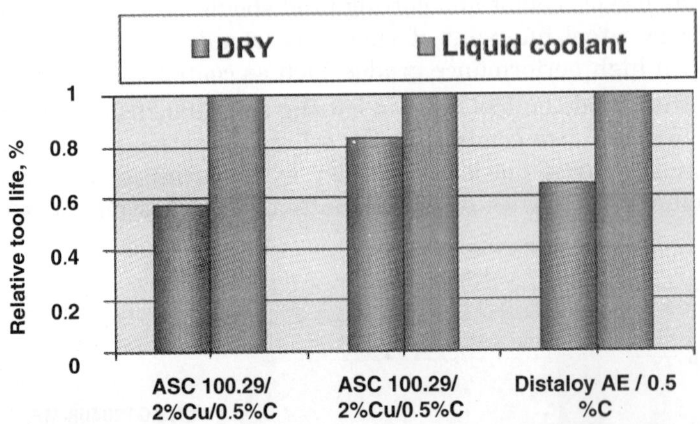

Fig. 23.14: Relative tool life by dry and liquid coolant cutting of three PM materials

23.18 TURNING TOOL RECOMMENDATIONS AND CUTTING DATA

The type of cutting tool used should be selected carefully. Coated cemented carbide (CVD and PVD) as well as Cermet and CBN (high CBN) tools can be used in P/M machining. CBN should preferably be used with the high abrasive alloys of Distaloy AE and Distaloy HP-1. Uncoated cemented carbide is seldom the best alternative unless a very low cutting speed is used. Recommended start values for selected turning inserts are shown in the Table 23.2 below.

Table 23.2: Recommended start values for selected turning inserts (dry cutting and 15 min. tool life)

P/M Type	Alloy element	Car-bon con-tent %	Spec. Cutting force, kc 0, N/mm^2	CVD-COATED		PVD-COATED	CERMET	UNCOAT	CBN
				GC3005	GC3025	GC1025	CT5015	H13A	CB7050
				Feed,mm/r					
				0.10–0.30	0.10–0.30	0.10–0.30	0.05–0.20	0.05–0.30	0.10
ASC 100.29 (Iron-based)		0	1900	220–230	195–200	195–200	115–125	115–120	
	0.45P 2Cu	0	2150	130–165	115–145	115–145	80–100	80–100	
	2Cu	0	2050	175–210	155–185	155–185	100–120	100–120	
	2Cu	0.25	2100	200–260	175–230	175–230	195–250	110–145	
	2Cu	0.5	2150	220–250	195–220	195–220	205–235	120–130	
		0.8	2300	170–235	150–205	150–205	170–230	100–130	
DISTALOYAE (Diffusion alloyed)	4Ni,1.5Cu, 0.5Mo	0	1900	200–210	175–185	175–185	150–160	95–100	
	4Ni,1.5Cu, 0.5Mo	0.25	2100	165–170	145–150	145–150	15–140	85–90	465
	4Ni,1.5Cu, 0.5Mo	0.5	2200	150–165	130–145	130–45	115–125	75–85	430
	4Ni,1.5Cu, 0.5Mo	0.8	2200	120–135	105–120	105–20	105–115	65–75	375
ASTALOY (Pre-alloyed 1.5Mo)	4Ni,2Cu	0.5	2250	175–200	140–160	140–60	160–80	95–105	
		0.5	2500	95–120	85–105	85–05	75–95	50–60	300

GC3005: First choice for high productive machining in continous and less demanding interrupted cutting.	GC1025: First choice or high security/versatility in continous and interrupted cutting.

Cutting speed recommendations depend upon the grade of tool material used and the type of P/M material being worked. Speeds can range from around 100 m/min for a uncoated grade, up to 500 m/min with CBN. Selection of feed rate is of utmost importance with P/M materials, more so than with most other workpiece materials in fact. With the exception of high speed applications, in particular CBN, most operations are limited by flank wear development. An increase of feed rate (*i.e.*, chip thickness) does not seem to have any negative effect on flank wear. It seems that flank wear is more dependent on the turned distance. Obviously the highest feed rate possible should be selected and unnecessarily low rates should be avoided. There are now insert on the market (*e.g.*, Sandvik Coromant WF and WM) which permit an increased feed rate while keeping surface roughness at a low level.

23.19 SELECTION OF DRILL TYPE

Several grades of high speed steel and cemented carbides have been vital in the development of machining tools for PM components. Some of these materials have themselves been produced with PM technology. For maximum productivity in terms of greater tool life, hole accuracy and other indicators of optimum performance, careful selection of drills, drill speeds and feed rates is essential.

Table 23.3 below provides recommendations of drill types for different materials and applications. The selection of drill type should be related to both the hardness of the material being machined and productivity. Each recommended drill application is rated as either excellent or acceptable. The speed and feed rates given are start values for dry cutting of PM materials with a density of approximately 7.0 g/cm^3. For each drill application, numerical data represent recommended cutting speeds for a tool life of 15 minutes. With each cutting speed a letter (T-Y) indicates recommended feed rate for the cutting operation.

Table 23.3: Recommended drill selection from eight drill types for a range of P/M materials. Cutting data is given for each recommended application

Selection of Drills	Standard Drills		Application, HSCo			Application Drills Solid Carbide		
Recommendations and cutting data	A120	A100	A520	A510	A511	R520	R522	R510
		A001	A521	A511		R550	R332	
			A524	A504				

■ Excellent for application
● Acceptable for applications

Example :
50=Peripheral speed m/min.mid range +/–10%

V=Feed per rev=See seperate Feed chart drills.

All recommendations dry when machining with cutting fluid increase speed by 10.50%

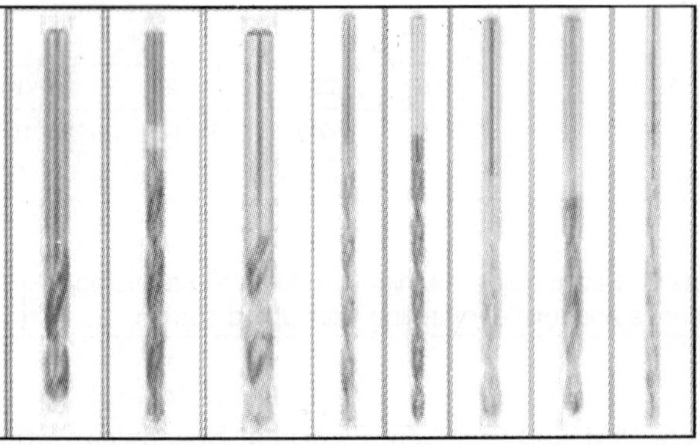

Material Groups	HSS 2.5xD Blue	HSS 4×D Blue	HSS/ HSCo 2.5×D Tin	HSS/H SCo 4×D TiN	HSCo 6×0 Atin	K10 25×0 Atin	K10 25×0 AtiN	K10 4×D Tin
ASC 100.29	■ 50V	■ 50V						
ASC 100.29+0.5%C	■ 64U	■ 64U						
ASC100.29+0.45%P	■ 38V	■ 38V						
ASC 100.29+0.45% P to 5%C	■ 38V	■ 38V						
ASC 100.29+0.45% P+0.5%C	■ 24U	■ 24U	■ 41Y	■ 41X	■ 41V			
ASC 100.29+2%Cu	■ 24V	■ 24V						
ASC 100.29+2%Cu+ 0.5%C	■ 50U	■ 50U	■ 97V	■ 87X	■ 57V			
ASC100.29+2% Cu+0.5%C	• 45V	• 45T	■ 78Y	■ 7X	■ 18V			
ASC100.29+2% Cu+0.8%C	• 25T	• 25T	■ 44Y	■ 44X	■ 44V			
Distaloy+4%Bu+ 1.5%Cu+0.5%Mo	■ 31U	■ 31U	■ 45X	■ 45W	■ 45U			
Distaloy+4%Ni+ 1.5%Cu+0.5% Mo+0.25%C	■ 31U	■ 31U	■ 48Y	■ 48X	■ 48V			
Distaloy+4%Ni+ 1.5%Cu+0.5%Mo+0.5%C	■ 20T	■ 20T	■ 31Y	■ 22Y	■ 27V	■ 83W	■ 83W	■ 60V
Distaloy +4%Ni+1.5%Cu +0.5%Mo+0.9%C			■ 20X	■ 27W	■ 20V	■ 83W	■ 83W	■ 60V
Astaloy+0.5% C (prealloyed 1.5 Mo)	■ 30U	■ 30U	■ 47Y	■ 27Y	■ 27U			
Astaloy+4%Ni+2%Cu+0.5%C. (prealloyed 1.5 Mo)			• 8X	• 8U	• 8U	■ 8U	■ 83W	■ 50V

23.20 CONCLUSIONS

Abrasive flank wear is the dominating wear mechanism in turning whereas drill length is considered as the primary factor for drill failure. Optimized feed rate increases tool life, especially with MnS and MnX additions. Cutting fluid is detrimental to turning but beneficial to drilling. Oil impregnation has a large effect for both turning and drilling. PVD-TiN coating of the HM inserts reduces the wear rate, CVD even further.

REFERENCES

1. M.C. Shaw. Metal Cutting Principles, Oxford, 1997.
2. A Thelin, "Verschleissmechanismen und Leistungen von Zerspanwerkzeugen" VDI Berichte, No. 762, 1989, pp. 111–126.
3. O.W Reen: 'The machinability of P/M materials', Modern Developments in Powder Metallurgy, Vol 10, pp. 431–452 (1977).
4. P.J. James: 'The machinability of Sintered Steels', PMI, Vol. 22, No. 6, 1990.
5. Höganäs Handbook for Sintered Components, Machining Guidelines, Höganäs AB, Sweden, 2004.
6. U. Engström: Machinability of Sintered Steel, Powder Metallurgy, Vol. 26, No. 3 1983, pp. 137–144.

❍❍

24

Effect of Machining on Work Hardening Characteristics of Powder Metallurgical Route Al/SiCp Metal Matrix Composites

*Uday A. Dabade, *V.V. Bhanuprasad[1], Suhas S. Joshi* and P. Ramakrishnan[2]*
**Deptt. of Mech. Engg., Indian Institute of Tech., Bombay, Mumbai, India*
[1]Defense Metallurgical Research Laboratory, Hyderabad, India
[2]Emeritus Professor, Indian Institute of Technology Bombay, Mumbai, India

Abstract

Highly abrasive nature of the reinforcement particles in Al/SiCp composites prepared by powder metallurgical route cause a number of difficulties in their machining. These include, excessive abrasive wear of cutting tools, sever damage and considerable work hardening of machined surfaces. All these aspects influence the characteristics of machined surfaces and thereby the performance of the components in real-life applications. In this paper, an analysis of work hardening behaviour of four different composites, prepared using powder metallurgical route, containing SiC reinforcements of two sizes (15 and 65 μm) and two volume fractions (20 and 30%) has been carried out. The experiments using Taguchi method-based L_{27} (3^{13}) orthogonal array have been carried out to investigate the effect of process and tool related parameters on work hardening behaviour of machined surfaces. It is observed that the work hardened zone below the machined surface called as altered material zone (AMZ) changes with change in size and volume fraction of reinforcement in composites. AMZ observed is about 150–210 mm for coarser and 90–150 μm for finer abrasive reinforced composites. Further the work hardened zone after machining is more sensitive to the size than volume fraction of abrasive reinforcement in composites. The finer abrasive reinforcement composites show higher (~ 60%) work hardening than the coarser reinforcement composites (~45%).

24.1 INTRODUCTION

In recent years, the need for less expensive structural materials that can provide an optimum level of performance has generated considerable research interest in the development and application of metal matrix composites (MMCs) [1-2]. MMC utilization provides significant benefits including performance (component lifetime, improved productivity), economic (energy savings) and environmental (lower noise levels and fewer air-borne emissions) [3-4]. The processing of composite material and presence of abrasive reinforcements profoundly affects the behaviour of the aluminum matrix composites during manufacturing and their subsequent use in service. Some of the effects include microstructural changes and induction of thermal stresses. These changes significantly alter the physical and mechanical properties of the composites [3].

It has been observed that the highly abrasive nature of the SiCp reinforcements in Al-matrix composites cause a number of difficulties in machining. These include, excessive abrasive wear of cutting tools, sever damage to the machined surfaces due to fracturing and dislodging of abrasive reinforcements, and considerable work hardening of machined surfaces. Numerous researchers have investigated the first aspect of the effect of abrasive reinforcements, but the second, and especially, the third effect has received very limited attention so far. There are only two references [2, 5] which refer to this topic and show that considerable work hardening of the machined surfaces occurs. Kishawy and Kannan [2] reported an altered material zone (AMZ) of about 45–60 μm while machining of alumina reinforced Al-matrix composites. The presence of abrasive reinforcement influences the solidification behaviour and age hardening characteristics of several aluminium alloys. During machining too, depending upon the matrix composition, size, volume fraction of the reinforcement, cutting tool geometry and machining conditions, the work hardening behaviour of machined surfaces changes. This indeed is the main focus of the investigation presented in this paper.

24.2 EXPERIMENTAL SET-UP AND PROCEDURE

The main objective of this research paper is to investigate the effect of change in size and volume fraction of abrasive reinforcement, tool geometry and processing parameters on work hardening of machined surfaces on Al/SiCp composites. The work hardening was measured in terms of variation in the micro-hardness of the layers below the machined surfaces. Accordingly, machining experiments using a Taguchi method-based orthogonal array L_{27} (3^{13}) [6] were performed on 2124 Al/SiCp composites having 220 (~ 65 μm) and 600 mesh size (~ 15 μm) with two volume fractions of SiCp reinforcement, 20 and 30%, processed by powder metallurgy route of vacuum hot pressing to full density composites have been used for the present investigations [7]. The chemical composition of the Al alloy-matrix is given in Table 24.1.

Table 24.1: Composition of 2124 Al alloy-matrix

Alloying element	Cu	Mg	Fe	Si	Cr	Ti	Zn	Mn	Al
wt. %	5.4	1.5	0.4	0.3	0.2	0.2	0.07	0.02	balance

Cutting tests were performed on CNC turning machine (EMCO/PCTURN Model 345-II) using CBN inserts with conventional (non-wiper, *NW*), wiper {*W1* (chamfer 20°×0.1mm width) and *W2* (chamfer 30°×0.1mm width)} insert geometry. Five independent variables at three levels each were used for the experimentation. They include two tool related parameters such as,

Table 24.2: Assessment of factors to L-27 orthogonal array and results of degree of work hardening

Expt. No.	Machining Parameters					Degree of Work Hardening (%)			
	Tool Nose Radius (mm)	Insert Geometry	Feed Rate (mm rev^{-1})	Cutting Speed (m min^{-1})	Depth of Cut (mm)	Al/SiCp 20/220	Al/SiCp 20/600	Al/SiCp 30/220	Al/SiCp 30/600
1	0.4	W1	0.05	40	0.2	50.91	65.56	43.33	54.00
2	0.4	W1	0.1	80	0.6	33.64	40.00	31.67	33.00
3	0.4	W1	0.2	120	1	20.91	32.22	27.50	27.00
4	0.4	W2	0.05	80	0.6	35.45	41.11	32.50	36.00
5	0.4	W2	0.1	120	1	31.82	38.89	30.00	32.00
6	0.4	W2	0.2	40	0.2	44.55	50.00	40.83	44.00
7	0.4	NW	0.05	120	1	40.00	44.44	36.67	39.00
8	0.4	NW	0.1	40	0.2	50.00	60.00	45.00	52.00
9	0.4	NW	0.2	80	0.6	18.18	27.78	16.67	21.00
10	0.8	W1	0.05	80	1	37.27	42.22	34.17	37.00
11	0.8	W1	0.1	120	0.2	41.82	46.67	35.83	41.00
12	0.8	W1	0.2	40	0.6	30.00	37.78	22.50	31.00
13	0.8	W2	0.05	120	0.2	43.64	47.78	39.17	42.00
14	0.8	W2	0.1	40	0.6	53.64	68.89	48.33	57.00
15	0.8	W2	0.2	80	1	29.09	36.67	25.83	30.00
16	0.8	NW	0.05	40	0.6	48.18	45.56	38.33	40.00
17	0.8	NW	0.1	80	1	30.91	38.89	29.17	32.00
18	0.8	NW	0.2	120	0.2	26.36	30.00	18.33	23.00
19	0.8	W1	0.05	120	0.6	40.91	52.22	38.33	48.00
20	0.8	W1	0.1	40	1	38.18	43.33	35.00	38.00
21	0.8	W1	0.2	80	0.2	36.36	42.22	33.33	37.00
22	0.8	W2	0.05	40	1	39.09	43.33	45.83	39.00
23	0.8	W2	0.1	80	0.2	34.55	41.11	32.50	35.00
24	0.8	W2	0.2	120	0.6	46.36	51.11	42.50	47.00
25	0.8	NW	0.05	80	0.2	49.09	55.56	44.17	50.00
26	0.8	NW	0.1	120	0.6	27.27	34.44	24.17	29.00
27	0.8	NW	0.2	40	1	21.82	31.11	20.83	25.00

tool nose radius (0.4, 0.8 and 0.8* mm), insert geometry (W1, W2 and NW) and three process related parameters *viz.*, feed rate (0.05, 0.1 and 0.2 mm rev^{-1}), cutting speed (40, 80 and 120 m min^{-1}), and depth of cut (0.2, 0.6 and 1 mm). The machining conditions for individual test were according to the L-27 orthogonal array are given in Table 24.2. Each turned cylindrical specimen was cut manually (using a hacksaw) into three parts (2 of ¼ size and one of ½ size). Two samples of ¼ size prepared were used for SE Microscopy and residual stress measurement on the machined surfaces respectively (these measurements results were presented elsewhere). The sample (of ½ size) has been used for the analysis of micro-hardness across the machined surfaces.

* Dummy level is selected which fulfills the requirement of L_{27} array and provides precise information about 0.8 mm TNR.

The ½ size samples cut from the machined specimens were cleaned ultrasonically with acetone and were encapsulated in an epoxy mold. Grinding of the specimens was then performed on water-proof silicon carbide papers mounted on rotary polishing wheels. The grinding sequence consisted of use of polishing papers of size 120, 240, 320, 400, 600, 800 and 1000 grit SiC. Coarser polishing was carried out on a Texmet pad (a napless cloth manufactured by buehler) using diamond paste of 6 μm grit size. Intermediate polishing was then performed using 6 μm diamond paste and a micro-cloth (medium nap synthetic suede cloth manufactured by Buehler). Commercially available mixture of distilled water and propylene glycol was used as a lubricant during polishing.

The final fine polishing was then performed by hand on a micro-cloth mounted on a glass plate using 0.04 μm colloidal silica suspension. The samples prepared for micro-hardness measurements are shown in Fig. 24.1. The measurement of micro-hardness variation across the machined surface on each specimen was carried out by taking indentations on the machined cross-sections at 30 to 240 μm beneath the machined surfaces at an interval of 30 μm, refer Fig. 24.2. The indentations were made in the Al-matrix away from the SiCp interface. Since it is known that the hardness of the Al-matrix varies when it is measured very close or away from the SiC particles.

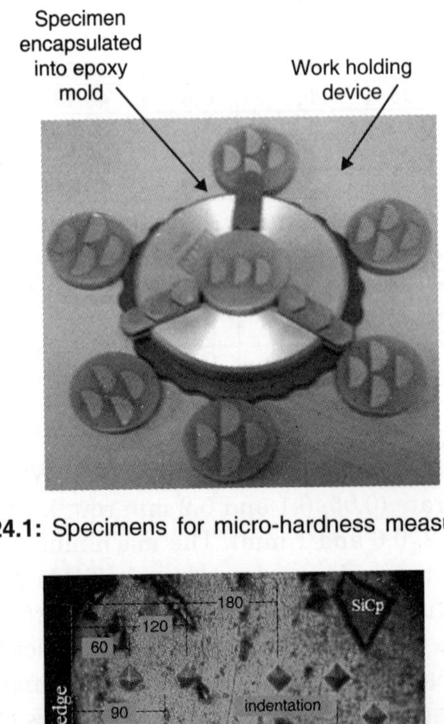

Fig. 24.1: Specimens for micro-hardness measurement

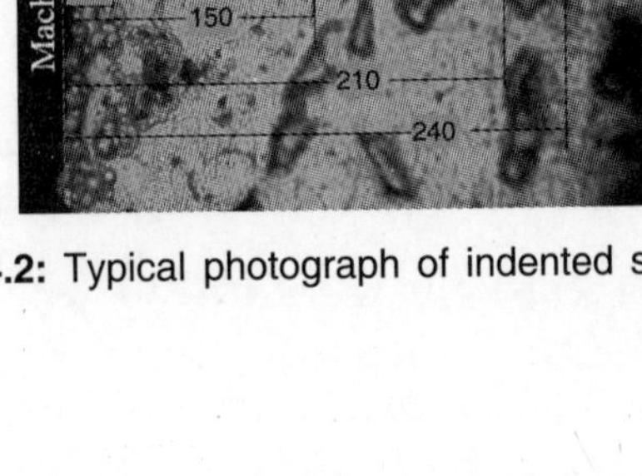

Fig. 24.2: Typical photograph of indented specimen

24.3 EXPERIMENTAL RESULTS AND DISCUSSIONS

The absolute magnitude of the micro-hardness measured along the cross-sections of the machined specimen at nine different locations between 30–240 μm below the machined surfaces at an interval of 30 μm is presented in Figs. 24.3(a-b) for Al/SiC/30p/220 and Al/SiC/30p/600 composites respectively (for two out of the four different composites used in this experimentation). It shows that the micro-hardness magnitude is much higher when measured close to the machined surface (30 μm beneath the machined surface). But the hardness decreases, away from the machined surface and reaches the bulk material hardness at a certain distance. This region, from the machined surface till the position at which it reaches the bulk hardness can be considered as a region affected by the machining process. Hence, it may be called as an altered material zone (AMZ).

24.3.1 Effect of Size and Volume Fraction of Abrasive Reinforcement on Micro-hardness Variation

The plots in Figs. 24.3(a-b) indicate that as the size and volume fraction of abrasive reinforcement in the composite material change, the extent of the AMZ changes, see Table 24.3. In the case of coarser abrasive reinforced composites, the AMZ is of the order of ~120–210 μm; whereas, for the finer abrasive reinforced composites it is ~90–150 μm. Thus, it is clear that the size of abrasive reinforcement determines the length of altered material zone than the volume fraction of reinforcement in composites that are considered in this experimentation.

An increase in the volume fraction of reinforcement in the Al-matrix reduces the mean average spacing between the abrasive SiC particles and they will act as barriers to the deformation of the Al-matrix material. This restriction increases the non-homogeneity in the plastic deformation and the tendency of work hardening. Therefore, the micro-hardness observed with higher volume fraction of reinforcement composite is more than the lower volume fraction of SiC reinforcement composite; see Table 24.3. This effect can also be evident from the elastic modulus of the composites. The composites with higher volume fraction of reinforcement have larger elastic moduli. Thus for the same reason Al/SiC/20p/220 composites has ~ 81 GPa and Al/SiC/30p/220 composites has ~ 99 GPa).

[a] Al/SiC/30p/220

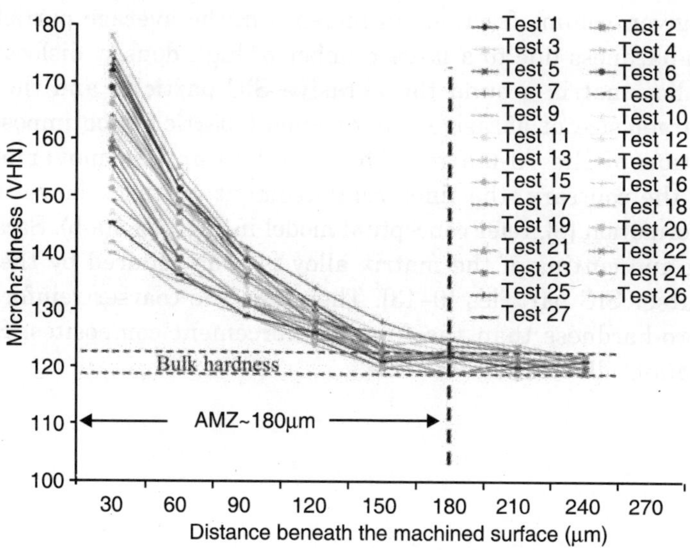

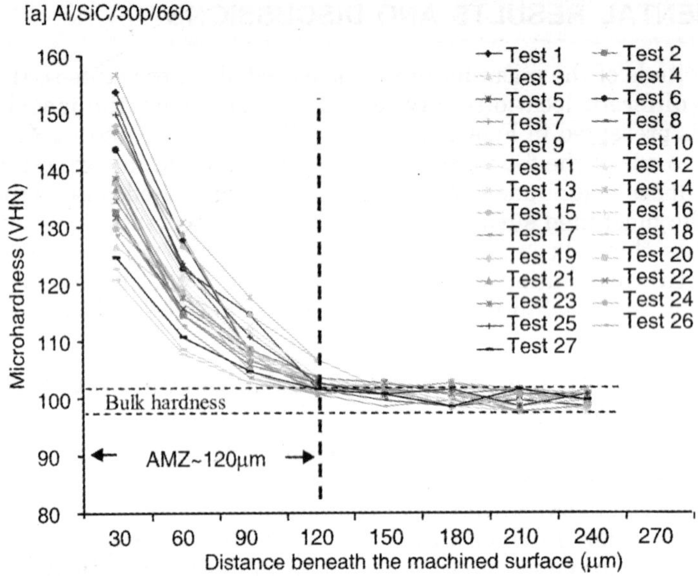

Fig. 24.3[a-b]: Effect of size of reinforcement on bulk hardness and altered material zone (AMZ)

Table 24.3: Magnitude of bulk micro-hardness and altered material zone

Material	Bulk hardness (VHN)	AMZ (m)
Al/SiC/30p/220	range of 119–122	~ 180–210
Al/SiC/20p/220	range of 108–112	~ 120–150
Al/SiC/30p/600	range of 99–102	~ 90–150
Al/SiC/20p/600	range of 88–92	~ 90–150

Again, for a given volume fraction, an increase in the average particle size, leads to an increase in micro-hardness due to a large number of high density dislocations that occur in the aluminium alloy matrix around the abrasive SiC particles and the matrix alloy gets hardened. The larger size of abrasive reinforcement particles too imposes barriers on the plastic deformation of the Al-matrix. They tend to oppose movement during plastic deformation. On the contrary, the finer reinforcement particles move with the Al-matrix during plastic deformation [8], refer conceptual model in Figs. 24.4(a-b). Secondly, the excessive plastic ploughing and cutting of the matrix alloy is been reduced by the presence of hard, undeformable coarser SiC particles [9–13]. Therefore, the coarser reinforcement composites show higher micro-hardness than the finer reinforcement composites. See the conceptual model in Fig. 24.4(a-b) illustrating this effect.

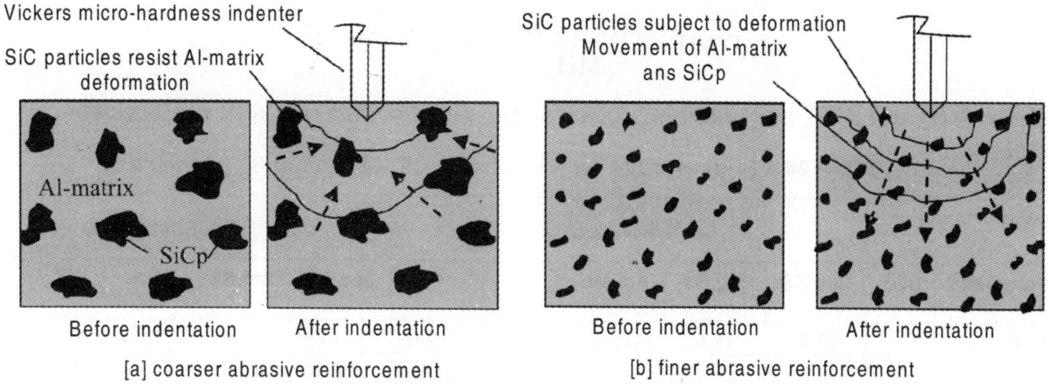

Fig. 24.4 [a-b]: Conceptual model illustrating the effect of size of reinforcement on micro-hardness indentations

24.3.2 Statistical Analysis of Microhardness

Though the micro-hardness measurements were carried out up-to 240 μm along the depth of the machined surface cross-sections, it was evident that after 90 μm depth, there is no significant variation in the micro-hardness; see Figs. 24.3 (a-b). However, the objective of the present work is to investigate the effect of work, tool and process related parameters on the work hardening characteristics of machined surfaces of Al/SiCp composites. Therefore, the analysis of variance of the micro-hardness data collected at 30 μm deep below the machined surface was carried out and is as shown in Table 24.4(a-d) for all the four types of composites used in this experimentation. It is observed that the feed rate, cutting speed and the depth of cut are the statistically significant parameters at 90% confidence level influencing the micro-hardness variation at 30 μm depth beneath the machined surfaces; refer P-values (< 0.1) in ANOVA Table 24.4 (a-d).

The analysis of means (AOM) plots for the three variables which influence the micro-hardness variation at 30 μm depth below the machined surface are shown in Fig. 24.5 (a-c). In each plot variation in micro-hardness is plotted against the cutting speed for all the composites chosen in this study. These plots indicate that the magnitude of micro-hardness on the machined surfaces is more at lower level of feed rate (0.05 mm rev^{-1}), cutting speed (40 m min^{-1}) and depth of cut (0.2 mm). Whereas, an increase in feed rate, cutting speed and depth of cut at higher levels reduces the micro-hardness beneath the machined surfaces.

An increase in the feed rate from 0.05 to 0.2 mm rev^{-1} and depth of cut from 0.2 mm to 1 mm deteriorates the quality of the machined surfaces and more boundary defects are generated by debonding/fracture/crushing of abrasive reinforcement particles in composites. It results in forming pit holes, cavities and cracks on the machined surfaces, which release the stresses. They induce tensile stresses on the machined surfaces and correspondingly reduce the micro-hardness along the machined surface cross-sections.

24.3.3 Degree of Work Hardening

The magnitude of degree of work hardening (DWH) calculated at 30 μm beneath the machined surface using Eqn. (1) [14] and is given in Table 24.1.

$$DWH = \frac{MH_s - MH_b}{MH_b} \times 100 \quad ...(1)$$

where, MH_s and MH_b are micro-hardness at the surface and the bulk respectively.

Table 24.4 [a-d]: ANOVA for micro-hardness variation at 30 μm depth

Micro-hardness 30 m		[a] Al/SiC/20p/220					[b] Al/SiC/20p/600				
Parameter	**DOF**	**Seq SS**	**Adj SS**	**Adj MS**	**F**	**P**	**Seq SS**	**Adj SS**	**Adj MS**	**F**	**P**
TNR	1	12.52	12.52	12.52	0.30	0.597	1.85	1.85	1.85	0.04	0.851
IG	2	146.74	88.11	44.06	1.05	0.388	122.30	61.04	30.52	0.61	0.563
FR (mm/rev)	2	841.41	817.44	408.72	9.79	**0.006**	474.30	472.15	236.07	4.74	**0.039**
CS (m/min)	2	387.63	387.63	193.81	4.64	**0.041**	334.30	334.30	167.15	3.36	**0.081**
DOC (mm)	2	522.74	522.74	26.37	6.26	**0.020**	347.63	347.63	173.81	3.49	**0.075**
TNR*IG	2	45.59	45.59	22.80	0.55	0.597	78.81	78.81	39.41	0.79	0.482
TNR*FR	2	25.59	25.59	12.80	0.31	0.744	13.48	13.48	6.74	0.14	0.875
IG*FR	4	545.04	545.04	136.26	3.26	**0.065**	359.70	359.70	89.93	1.81	0.212
Error	9	375.93	375.93	41.77			447.93	447.93	49.77		
Total	26	2903.19					2180.3				
		R^2 = 87.05% ; R^2 (adj) =62.59%					R^2 = 79.46% ; R^2 (adj) = 40.65%				
Micro-hardness 30 m		[c] Al/SiC/30p/220					[d] Al/SiC/30p/600				
Parameter	**DOF**	**Seq SS**	**Adj SS**	**Adj MS**	**F**	**P**	**SeqSS**	**Adj SS**	**Adj MS**	**F**	**P**
TNR	1	0.00	0.00	0.00	0.00	1.000	0.46	0.46	0.46	0.01	0.927
IG	2	330.89	191.59	95.80	1.53	0.268	151.1	77.15	38.57	0.74	0.504
FR (mm/rev)	2	881.56	713.04	356.52	5.69	**0.025**	570.0	536.9	268.5	5.15	**0.032**
CS (m/min)	2	320.67	320.67	160.33	2.56	0.132	287.2	287.2	143.6	2.75	**0.110**
DOC (mm)	2	200.67	200.67	100.33	1.60	**0.054**	347.6	347.6	173.8	3.33	**0.083**
TNR*IG	2	100.78	100.78	50.39	0.80	0.477	72.70	72.70	36.35	0.70	0.523
TNR*FR	2	27.11	27.11	13.56	0.22	0.809	4.48	4.48	2.24	0.04	0.958
IG*FR	4	393.56	393.56	98.39	1.57	0.263	409.9	409.9	102.5	1.96	0.184
Error	9	563.44	563.44	62.60			469.5	469.5	52.17		
Total	26	2818.67					2313				
		R^2 = 80.01% ; R^2 (adj) = 42.25%					R^2 = 79.70% ; R^2 (adj) = 41.36%				
Bold value indicates that the corresponding parameters are significance at 90% confidence level											

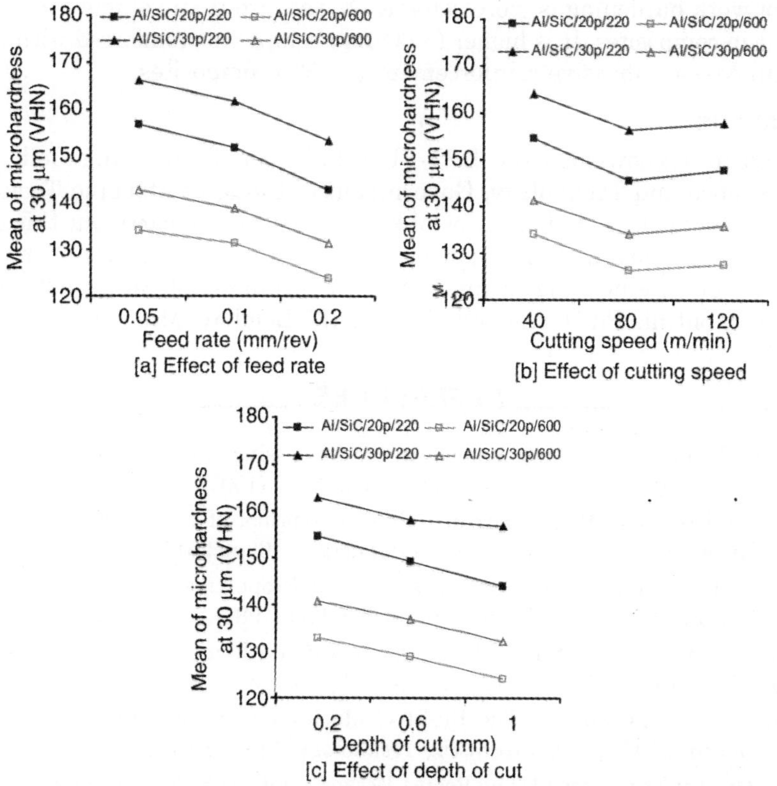

Fig. 24.5 [a-c]: Effect of process parameters on micro-hardness variation

It is observed that the degree of work hardening is higher with finer abrasive reinforcement composites (about 60%) than coarser abrasive reinforcement composites (about 45%), refer Table 24.1. This is attributed to the uniform deformation of matrix material in finer abrasive reinforcement composites. Under the identical machining condition, the average plastic deformation of composites reinforced by finer abrasive particles is larger than that of composites reinforced by coarser abrasive particles. However, the former deforms more uniformly, which results in a slight tool wear and lower cutting temperature, thus, there is no recast structure in its surface layer, and the machined surface layer is work-hardened. Similar results were reported by Quan and Bangyan [5] while machining of Al356/SiC/15p composites having reinforcement size of 14, 42 and 85 μm and Yan et al. [15] in their investigations.

24.4 CONCLUSIONS

Based on the experimental investigations following conclusions can be drawn.

- The depth of work hardened or altered material zone in the machined surfaces changes with size of abrasive reinforcement in composites. In the case of coarser composites, the AMZ is in the order of ~ 120–210 μm, and for the finer reinforced composites, it is ~ 90–150 μm.
- Feed rate, cutting speed and depth of cut are the significant parameters, which influence the micro-hardness variation beneath the machined surface during machining of Al/SiCp composites.

- The degree of work hardening is more sensitive to the size than volume fraction of abrasive reinforcement in composites. It is higher (~ 60%) in composites reinforced with finer abrasive particles than coarser abrasive reinforcement (~ 45%) composites.

Acknowledgements

The authors wish to acknowledge the DST-FIST Precision Prototyping Cell sponsored by Department of Science and Technology, Government of India, at Machine Tools Laboratory, Indian Institute of Technology, Bombay, for providing facilities to carry out this experimental work. The author Dr. Uday Dabade wish to acknowledge the support of management of Walchand College of Engineering, Sangli, and Government of Maharashtra, for giving opportunity to carry out his Ph.D. research work at IIT Bombay, Mumbai under QIP scheme.

REFERENCES

1. Ejiofor, J.U. and Reddy, R.G., "Developments in the Processing and Properties of Particulate Al-Si Composites", *Journal of Material Science*, 49(11), pp.31–37 (1997).
2. Kannan, S. and Kishawy, H.A., "Surface Characteristics of Machined Aluminium MMCs, *International Journal of Machine Tools and Manufacture*, 46, pp.2017–2025 (2006).
3. Surappa, M.K. and Rohatgi, P.K., "Preparation and Properties of Al-alloy Ceramic Particle Composites", *Journal of Material Science*, 16, pp.983–993 (1981).
4. Chambers, A.R., and Stephens, S.E., "Machining of Al-5 Mg Reinforced with 5 vol% Saffil and 15 vol% SiC Fibres", *Journal of Material Science and Engineering A* 135, pp.287–290 (1990).
5. Yanming Quan, and Bangyan Ye, "The Effect of Machining on the Surface Properties of SiC/Al Composites", *Journal of Materials Processing Technology*, 138, pp.464–467 (2003).
6. Phadke, M.L., "Quality Engineering Using Robust Design", *Prentice Hall Publication*, New Jersey, 1989.
7. Bhanuprasad. V.V., Bhat. B.V.R., Somaraju. K. and Ramakrishnan. P., "Development of MMC Connecting Rods for Automotive Applications", Powder Metallurgy in Automotive Applications-II, Eds. T. R. Rama Mohan and P. Ramakrishnan. Oxford & IBH Publishing Co. Pvt. Ltd., New Delhi, pp.257–266 (2002).
8. Quan Yanming and Zhou Zehua, "Tool Wear and its Mechanism for Cutting SiC Particle-Reinforced Aluminium Matrix Composites" *Journal of Materials Processing Technology*, 100, pp.194–199 (2000).
9. Lee, H.L., Chan, S.L. and Lu, W.H., "Abrasive Wear of Powder Metallurgy Aluminium Alloy 6061-SiCp composites", *Wear*, 59(2), 223–231 (1992).
10. Zamzam, M.A., "Abrasive wear of Aluminium Matrix Composites", Metall, 45, pp.250–254 (1991).
11. Prasad, S.V. and Rohatgi, P.K., "Tribological Properties of Al-alloy Particle Composites", JOM, 39(11), pp.22–26 (1987).
12. Alpas, A.T. and Embury, J.D., "Sliding and Abrasive Wear Behaviour of an Aluminium (2014)-SiC Particle Reinforced Composite", *Scripta Materialia*, 24, pp.931–935 (1990).
13. Lin, S.J. and Liu, K.S., "Effect of Aging on Abrasion Rate in an Al-Zn-Mg-SiC Composite", *Wear*, 121, pp.1–14 (1988).
14. Liu, C.R. and Barash, M.M., "The Mechanical State of the Sub Layer of a Surface Generated by Chip-Removal Process Part 1: Cutting with a Sharp Tool", *Transactions of ASME, Journal of Engineering for Industry*, pp.1192–1201 (1976).
15. Yan Y.W., Geng L. and Li A. B., "Experimental and Numerical Studies of the Effect of Particle Size on the Deformation Behaviour of the MMCs" *Material Science and Engineering A*, 448, pp.315–325 (2007).

❍❍

Index